Chemical Deterioration of Proteins

Chemical Deterioration of Proteins

John R. Whitaker, EDITOR

University of California, Davis

Masao Fujimaki, EDITOR

Ochanomizu University

Based on a symposium

sponsored by the Division of

Agricultural and Food Chemistry

at the ACS/CSJ Chemical Congress,

Honolulu, Hawaii,

April 4–5, 1979.

ACS SYMPOSIUM SERIES **123**

AMERICAN CHEMICAL SOCIETY

WASHINGTON, D. C.   1980

Library of Congress CIP Data

Main entry under title:
Chemical deterioration of proteins.

(ACS symposium series; 123 ISSN 0097-6156)

Includes bibliographies and index.

1. Proteins in human nutrition—Congresses. 2.
Proteins—Deterioration—Congresses.
I. Whitaker, John R. II. Fujimaki, Masao. III.
American Chemical Society. Division of Agricultural
and Food Chemistry. IV. Series: American Chemical
Society. ACS symposium series; 123.

TX553.P7C44   664   79-27212
ISBN 0-8412-0543-4 ACSMC8 123 1–268 1980

ACS Symposium Series

M. Joan Comstock, *Series Editor*

FOREWORD

The ACS SYMPOSIUM SERIES was founded in 1974 to provide a medium for publishing symposia quickly in book form. The format of the Series parallels that of the continuing ADVANCES IN CHEMISTRY SERIES except that in order to save time the papers are not typeset but are reproduced as they are submitted by the authors in camera-ready form. Papers are reviewed under the supervision of the Editors with the assistance of the Series Advisory Board and are selected to maintain the integrity of the symposia; however, verbatim reproductions of previously published papers are not accepted. Both reviews and reports of research are acceptable since symposia may embrace both types of presentation.

CONTENTS

Preface ix

1. Overview on the Chemical Deteriorative Changes of Proteins
   and Their Consequences 1
   Robert E. Feeney

2. Posttranslational Chemical Modification of Proteins 49
   Rosa Uy and Finn Wold

3. Chemical Changes in Elastin as a Function of Maturation 63
   Robert B. Rucker and Michael Lefevre

4. Photooxidative Damage to Mammalian Cells and Proteins by
   Visible Light 83
   L. Packer and E. W. Kellogg, III

5. Chemical Deterioration of Muscle Proteins During Frozen Storage . 95
   Juichiro J. Matsumoto

6. Preservation of Enzymes by Conjugation with Dextran 125
   J. John Marshall

7. Changes Occurring in Proteins in Alkaline Solution 145
   John R. Whitaker

8. Amino Acid Racemization in Alkali-Treated Food Proteins—
   Chemistry, Toxicology, and Nutritional Consequences 165
   Patricia M. Masters and Mendel Friedman

9. Deterioration of Food Proteins by Binding Unwanted Compounds
   Such as Flavors, Lipids and Pigments 195
   Soichi Arai

10. Deteriorative Changes of Proteins During Soybean Food
    Processing and Their Use in Foods 211
    Danji Fukushima

11. Suicide Enzyme Inactivators 241
    Brian W. Metcalf

Index 255

PREFACE

Protein molecules are synthesized rapidly (3–5 minutes) in vivo with a high degree of precision. The error level in the incorporation of specific amino acids into a growing polypeptidyl chain to give the primary sequence of a specific protein is estimated at about one error in every 10^4 to 10^5 amino acids incorporated. Unlike carbohydrates, every molecule of a given protein is identical in molecular weight, amino acid sequence, and secondary structure. When the proteins are released from the ribosomes they immediately are confronted with a hostile environment. Some proteins survive in the environment for only a few minutes while others last several years. For example, the half life of ornithine decarboxylase normally is about 11 minutes, while that of elastin is not readily measurable.

Either while on the ribosome or immediately after release, many proteins undergo posttranslational modifications in which certain amino acid side chains are specifically modified (for example, conversion of proline and lysine to hydroxyproline and hydroxylysine) or derivatized (for example, glycosylated, acylated, or phosphorylated). When secreted from the cell, the cysteinyl residues are often oxidized to form disulfide bonds, or cross-links. Proteins are often synthesized in precursor, inactive forms and must undergo limited proteolysis in order to become biologically active (for example, certain zymogens of hydrolytic enzymes, proinsulin, proelastin). Collagen and elastin molecules undergo extensive crosslinking via the action of lysyl oxidase and subsequent nonenzymatic condensation reactions of the allysine formed with amino and other nucleophilic groups of the same protein. While some 20 different amino acids are incorporated into proteins via translation, more than 150 different amino acids have been found in proteins long after biosynthesis.

Following translation into the polypeptidyl chain, proteins undergo not only specific chemical modifications but also many nonspecific modifications. Some modifications are the result of continual exposure of the proteins to the potential action of proteolytic enzymes. The level of a specific protein in vivo is the result of a balance achieved between the rate of biosynthesis of that protein and its rate of degradation by proteolytic enzymes. In part, the rate of degradation of a protein is a function of the relative levels of native (N) and reversibly (R) denatured protein ($N \rightleftharpoons R$) under a given set of conditions. Nutritional state and health of the organism, the extent of posttranslational modification, and environmental damage to the proteins will influence their rates of turnover (probably by influencing the equilibrium between N and R). Other nonspecific protein modifications occur secondarily as a result of free radical reactions initiated by light, oxygen, ozone, hydrogen peroxide, nitrous acid, and other reagents. The rates of these reactions, and others, are enhanced by the presence of polyunsaturated lipids, especially at low levels of vitamin E and other antioxidants, and at low levels of glutathione peroxidase. A few proteins (polyphenol oxidase, ascorbic acid oxidase, etc.) undergo substrate or product modification during the reaction.

Proteins may also be subjected to harsh conditions leading to physical and chemical modification during the storage and processing of food materials. A few examples include the interaction of benzoquinones with proteins, the covalent interaction of reducing sugars with the amino groups of proteins during storage or processing (Maillard or nonenzymatic browning), interaction of proteins with pigments and with added dyes, and the many changes which occur when proteins are treated with alkali in order to solubilize them for texturization. Extensive changes in proteins may occur even during frozen storage (toughening and loss of water binding properties of fish muscle, for example).

Proteins may also be intentionally modified, as in cooking, in order to increase their digestibility with proteolytic enzymes as a result of denaturation, to destroy certain antinutritional proteins (toxins, enzyme inhibitors), and to improve the flavor and texture. We already mentioned the use of alkali in solubilization and texturization of proteins and the browning of bread. Modification of protein is induced in dough formation in order to improve the extensibility and carbon dioxide–holding properties of the dough. Proteins may also be deliberately treated with chemical reagents in order to improve the nutritional quality (by covalent incorporation of limiting essential amino acids), prevention of the Maillard reaction by protecting the amino groups, and modification of functional properties (solubility, and whipping, foaming, and emulsifying properties).

These reactions, whether occurring in vivo or in vitro, unintentionally or deliberately, result in chemical and physical deterioration of the proteins. As noted by the examples above, chemical deterioration may be essential or nonessential, beneficial or detrimental. The same reaction can be detrimental in one case and beneficial in another. The purpose of this volume is to explore these reactions in detail in order to maximize their benefits in the processing and formulation of our food.

JOHN R. WHITAKER
University of California
Davis, California

MASAO FUJIMAKI
Ochanomizu University
Tokyo, Japan

September 1979

Overview on the Chemical Deteriorative Changes of Proteins and Their Consequences

ROBERT E. FEENEY

Department of Food Science and Technology, University of California, Davis, CA 95616

The deteriorations and the deteriorative reactions of proteins have been studied by scientists in many different fields for many centuries. In order to give proper tribute to the almost ancient importance of proteins, it would be necessary to summarize the history of agriculture, medicine, food processing, and much of industry. Scientists and technologists have long recognized both the adverse and beneficial facets of deteriorative changes in proteins.

Putrefactive and coagulative processes might be considered two of the oldest and perhaps most investigated areas of protein chemistry. The disgustingly bad odors from the breakdown products of sulfur amino acids were, by perforce, of everlasting concern while the coagulative processes were probably part of ancient art, certainly of cooking, as well as of industrial and medical technologies. Most likely it is the ever obtrusive phenomenon of protein coagulation that even today may be responsible for the difficulty in differentiating between the initial, more delicate, steps of protein denaturation and the extensively devastating processes surrounding coagulations resulting from extreme treatments such as boiling.

The isolation, preservation, and analysis of proteins were among the primary areas of protein chemistry until the early 20th century. In nearly every step of isolating proteins, workers encountered the problem of preventing deteriorative reactions and, as a consequence, began to study the deteriorative reactions themselves. Many of these earlier studies of deteriorative reactions have now been described in quantitative chemical terms, but many still elude the efforts of current workers using modern techniques.

The immensity of this subject at first made it seem that an overview could only be done one of two ways, 1) essentially a many-page outline of the deteriorations, or 2) a selection of two or three deteriorative reactions and their coverage in a comparative and illustrative way. There were suggestions from several sources that a more illustrative coverage could be based on the

0-8412-0543-4/80/47-123-001$11.75/0

author's long interest in deteriorative reactions. A third
approach has therefore been taken: a general coverage, with more
details for those studies with which the author has firsthand
knowledge. Omissions of citations to many areas are a consequence of these selections by the author as well as of the large
amount of material. It is hoped that the many fine articles in
this volume will compensate for these omissions.

The Widespread Occurrence of Protein Deteriorations

Deteriorative reactions of proteins are important in almost
every biological system, whether alive or dead. Until recently,
most studies dealt with those deteriorative processes occurring
on the death of a system or in its storage or handling, such as
in food products. More recently, in an ever-increasing volume,
publications have appeared citing studies of naturally occurring
deteriorative changes, both beneficial and detrimental.
Two biologically related processes which have received attention for many years are the clottings of milk and blood.
Blood clotting, an exceedingly complex cascading system involving
numerous activations of zymogens, and subsequent amplification of
products, is a series of syntheses via degradations, in each case
involving breakdown of a precursor. Many other biological processes are today under stringent and extensive study. These
processes include the activations and inactivations that can
occur by the additions or removals of such substances as phosphate groups, carbohydrates, or fragments of peptides, as well as
by the limited scission (clipping) of the peptide chains of
proteins.

Denaturation

The complex structure of proteins and the many different
kinds of protein structures are responsible for the different responses of proteins to environmental stresses. Denaturation is a
term which has been used with many different meanings. In its
broadest sense it means "away from the native state". In its
more strict thermodynamic sense, it is defined as "change from an
ordered to a disordered state - an increase in entropy". A more
practical and everyday working definition is "the change in protein structure that is not accompanied by, or caused by, any making or breaking of covalent bonds". Denaturation is therefore a
physical process rather than a chemical one, although it is easily induced by chemical reagents, and consequently might be
omitted from a discussion of chemical deterioration of proteins.
Any discussion of protein deterioration must, however, include at
least a limited discussion of denaturation because it is one of
the most important deteriorative reactions of proteins, and it is
necessary to differentiate between denaturation and chemical deteriorations. Denaturation should thus always be considered.

Denaturation is almost always applied to changes in globular proteins. Fibrous proteins, such as hair, can obviously be made to change their physical state by physical means, and such changes might therefore be called denaturation, but these changes are usually not described as denaturation. In this discussion we will restrict the coverage to globular proteins.

Denaturation was early observed to be a reversible process. Indeed, Anson (1) observed 35 years ago that hemoglobin could be heat denatured in a variety of ways and could be converted back to a state which had all the characteristics of its original native state, as determined by methods available at that time. Almost all studies of protein denaturation now revolve around not only the denaturation itself, but also its renaturation; perhaps renaturation is a more interesting and provocative field for modern research.

There have been many recent reviews of denaturation and renaturation (2,3) and the many related theoretical areas, such as the effects of amino acid composition and microenvironment on protein structure (4), the empirical prediction of protein conformation (5), and the experimental and theoretical aspects of protein folding (6).

The processes of both denaturation and renaturation are intimately related to the structures of native proteins. Alpha helices and β-pleated sheets constitute the main structures in most all native proteins. How the helices and sheets pack together depends on the geometrical characteristics of their surfaces. Contacts may exist on all sides and, although nonpolar (hydrophobic) side chains are buried inside, water may be present in crevices as well as in pools on the surface. It is through the disarrangement and rearrangement of all these, and more, structures that the pathways of denaturation and renaturation are directed.

Modern theories of protein structure state that the amino acid sequence of the protein dictates the final conformation of the protein. If this were so, exposing the protein to a denaturing environment should not destroy the dictatorial powers of the primary structure; consequently, placing the protein back into its former environment should allow the protein to resume its native structure. This simple concept implies that the native form of the protein is at its lowest free energy state. This is illustrated in Figure 1. This simple thermodynamic picture, however, is not completely in line with observed facts. There appear to be "structures within structures" in the protein which could introduce kinetic pathways that might put the protein in a different final state than that represented by the minimal free energy. These "structures within structures" have been termed LINCS (local independently nucleated continuous segments) (7). Protein folding would then be like that shown in Figure 2, where the protein does not roll up into its original globular ball shape in one process, but rather assumes small areas of nativity, which then assume the final native state.

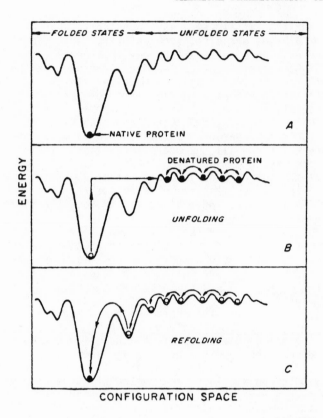

Figure 1. *Highly schematic diagrams of the energy of a protein molecule as a function of chain conformation (4)*

I. Polypeptide chain
   as synthesized

II. Local folding as
    dictated by local
    sequence - formation
    of LINCS

III. Tertiary folding of
     chain at inter-LINC
     joints to minimize
     free energy of LINC
     structure

Figure 2. *Protein folding in terms of the LINCS hypothesis (4)*

It must be emphasized that denaturations are also possible by many different routes, and the intermediate structures through which the protein would pass in assuming a completely denatured state would be different with different denaturing conditions.
Denaturation is a highly cooperative process. This is easily seen from the large values for the transition state denaturation constants for proteins (Table I). Very large $\Delta S^{\ddagger}$ values, the entropic term, are seen - the early phase of denaturation is

Table I. Transition State Denaturation Constants for
Various Proteins ($\underline{2}$)

Protein	$\Delta H^{\ddagger}$ (cal/mole)	$\Delta S^{\ddagger}$ (e.u.)[a]	$\Delta F^{\ddagger}(25°C)$ (cal/mole)
Trypsin	40,200	44.7	26,900
Pepsin	55,600	113.3	21,800
Hemoglobin	75,600	152.7	30,100
Egg albumin	132,000	315.7	37,900
Peroxidase, milk	185,300	466.0	46,400

[a]In cal/mole/degree at 25°C.

Avi Publishing Company

a highly cooperative process; later phases could be considered cascading processes. Once the critical temperature range for denaturation is reached, slightly increased severity of the conditions, such as a small increase in temperature, greatly increases the speed of denaturation. It is probably for this reason that proteins are considered so sensitive to denaturation in commercial processing procedures. There is a delicate temperature range, dependent on other environmental conditions as well, beyond which further treatment may result in undesirable denatured products, frequently ending in coagulums.

In common with most laboratories engaged in fundamental research on proteins, our laboratory has studied the denaturation and renaturation of proteins. Many of these studies have been with the two related homologous iron-binding proteins, human serum transferrin and chicken ovotransferrin. Earlier studies showed that on the binding of iron these proteins were greatly stabilized against denaturation by a variety of environmental stresses as well as to chemical scission of their disulfide bonds and to hydrolysis by proteolytic enzymes ($\underline{8},\underline{9}$). Such a seemingly simple question as to why these iron complexes, as well as some other proteins, are much more stable than others is still impossible to answer with presently available information.

Our laboratory has recently been concerned with the denaturation of chicken egg-white ovotransferrin by acid or urea

and the renaturation processes from each of these treatments.
When ovotransferrin is denatured by acid or urea, there is an extensive change in shape, resulting in decreases in both the sedi-
mentation velocity coefficient and the diffusion constant; these
are accompanied by a corresponding increase in the viscosity
(10). Ovotransferrin, in common with its homologous protein,
serum transferrin, has two separate iron-binding sites and is
reported to be the product of gene duplication (11), suggesting,
in present day terms, that it may consist of two domains. The
physical changes observed on denaturation could be interpreted
as being due to an unfolding of the molecule, a change which
would perhaps be in agreement with a model of two separate domains unfolding at, or near, some possible connecting link. An
equal possibility, however, would be a simple swelling of the
molecule.

Our laboratory recently reported on a study of the conformational properties of ovotransferrin, its denatured form (by
treatment with acid or urea) and its renatured form. The samples
were denatured in 7.2 $\underline{M}$ urea or in acidic (pH 3) conditions for
periods up to a few hours. Samples were renatured by dilution
and adjustment of the pH to neutrality, or by simple dilution of
the urea. Combined data from quasi-elastic light scattering and
transient electric birefringence were used to estimate the molecular dimensions under the various conditions. Analytical
ultracentrifugation was used to determine the changes in sedi-
mentation coefficient, and changes in helicity were calculated
from circular dichroism data. The course of renaturation as
measured by the increase in diffusion during renaturation of acid
denatured ovotransferrin is seen in Figure 3. Structural changes
from circular dichroism data of the native, urea-denatured, and
renatured sample are seen in Figure 4. A summary of the data and
calculations from the urea denaturation studies is in Table II.
The conclusion from these data was that, on denaturation, the
protein assumed a more expanded globular form than the native
sample; in other words, it swelled, rather than unfolded.

Chemical Reactions of Amino Acids of Concern in Deteriorations

Approximately 150 different amino acid residues have been
reported in proteins (15). At least half of these could undergo
chemical deteriorations under the conditions of stress usually
encountered. Many of these deteriorative reactions involve
hydrolytic scissions, not only of peptide bonds but of the many
different nonprotein substances added covalently to proteins
postribosomally. These susceptible side chain groups are indole,
phenoxy, thioether, amino, imidazole, sulfhydryl, and derivatives
of serine and threonine (such as O-glycosyl or O-phosphoryl), the
disulfides of cystine, and, of course, the amides (such as
asparagine and glutamine). With strong acid or alkali, other
residues, such as serine and threonine, also are less stable.

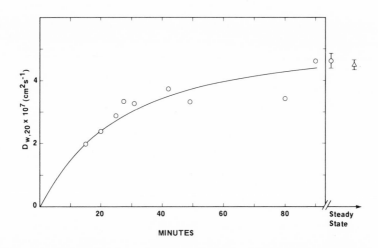

Figure 3. *Time development of the renaturation process of acid-denatured ovotransferrin. Concentration of ovotransferrin in the denatured state (pH 3) was approximately 10 mg/ml. The sample was diluted 10:1 in Tris buffer at pH 7.8. Note comparison values of D_t for steady-state native ($\triangle$), and renatured ($\bigcirc$) samples (12).*

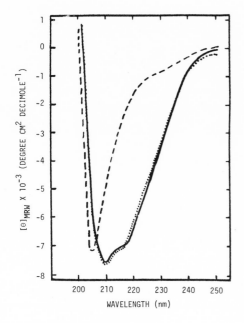

Figure 4. *Circular dichroism spectra from 200 to 250 nm for ovotransferrin. Mean residue weight of 112 is used. Native, c = 1.04 mg/mL (———); 7.2M urea-denatured sample, c = 1.04 mg/mL (– – –); and renatured sample, c = 0.83 mg/mL. (· · ·) (12).*

Table II. Denaturation and Renaturation of Ovotransferrin: Urea Denaturation (12)

Sample	Quasi-elastic Light Scattering $<D_t>_z \times 10^7$ cm^2s^{-1}	Conversion[a] $<D_t> \times 10^7$ cm^2s^{-1}	Transient Electric Birefringence $<\tau>$ µs	Viscosity Corrected to w, 20°C $<D_R>_w \times 10^{-6}$ s^{-1}[b]	Sense of Biref. δ_B
Native	4.31 ±0.3	6.14±0.43	0.82±0.14	2.17±0.32	−
Denatured 7.2 M urea	2.70±0.15	3.86±0.21	0.81±0.11	0.72±0.08	+
Renatured	4.21±0.30	6.00±0.43	0.93±0.07	1.91±0.13	−

[a]The dispersion factor, $\delta = 0.1$ in the following equation: $<D>_w \simeq D_z/(1-3\delta)$ were D_w and D_z are the weight- and z-averaged diffusion coefficients, respectively.

[b]Weight averaging is obtained directly by transient electric birefringence.

Table II. Denaturation and Renaturation of Ovotransferrin: Urea Denaturation (12) (continued)

Sample	Perrin eq[c] Prolate Ellipsoid Å a	b	ρ	Solvation Factor[d] $\delta_1 \frac{mg}{mg\ soln}$	Circular Dichroism[f] α%	β%	R.C.%	Fe^{+3} Binding[g] %
Native	68	21	0.31	0.28	11 (11)	32	57	86
Denatured 7.2 M urea	84	42	0.50	4.16[e]	6 (0)	19	75	< 5 (0)
Renatured	67	22	0.33	0.36	12 (10)	29	59	82

[c] Oblate ellipsoid values were disc-like, incompatible with other reported results.

[d] $\bar{v}$ = 0.73, molecular weight = 77,000 values used in determining δ_1.

[e] Reason for large value is not known at this time.

[f] Values in parenthesis are calculated from method of Chen and Yang (13), other values by method of Greenfield and Fasman (14).

[g] Values are calculated from direct color determinations on solutions. True values should be higher.

But even the relatively resistant residues are attacked by free
radicals.
When proteins are deliberately treated with chemicals in
order to derivatize them, the reaction conditions may also cause
chemical deteriorative side reactions. Some of the more common
ones are listed in Table III. Inspection of Table III shows that
many of these effects are those found in deteriorative reactions

Table III. Possible Chemical Side Reactions during
Protein Modification

Groups	Treatment	Effects
Peptide bonds	Alkaline pH	Hydrolysis
	Acidic pH	N→O acyl shift
	Alk, heat	Racemization
Thiol groups	Oxidation	-S-S-, acids
Disulfide bonds	Reduction	-SH, mispairing
	Alkaline pH	Hydrolysis, β elimination
Methionyl groups	Oxidation	Oxy sulfurs
Amide groups	Alkaline pH	Hydrolysis
O-Glycosyl	Alkaline pH	β Elimination
O-Phosphoryl	Alkaline pH	β Elimination

Biochemistry

resulting from other treatments; in other words, they are, in
some cases, environmentally produced rather than a direct result
of the chemical procedure. An example of a variety of reactions
caused by a relatively mild reagent are those with hydrogen peroxide (Figure 5). Hydrogen peroxide readily reacts with three
different side chain groups under mild conditions, and the extent
of the reaction is influenced by the presence of other substances, such as organic acids, that can form more active oxi-
dizing agents.
    As is the case with some chemical changes occurring in bio-
logical systems, such as the blood-clotting cascade system, deteriorative reactions considered to have a beneficial effect are
found in foods. For example, the Maillard reaction (17,18) is
used to produce flavors and colors in such foods as beverages and
baked goods. Heat treatment (involving denaturation) has been
found to increase the nutritional value of raw soybean meal by

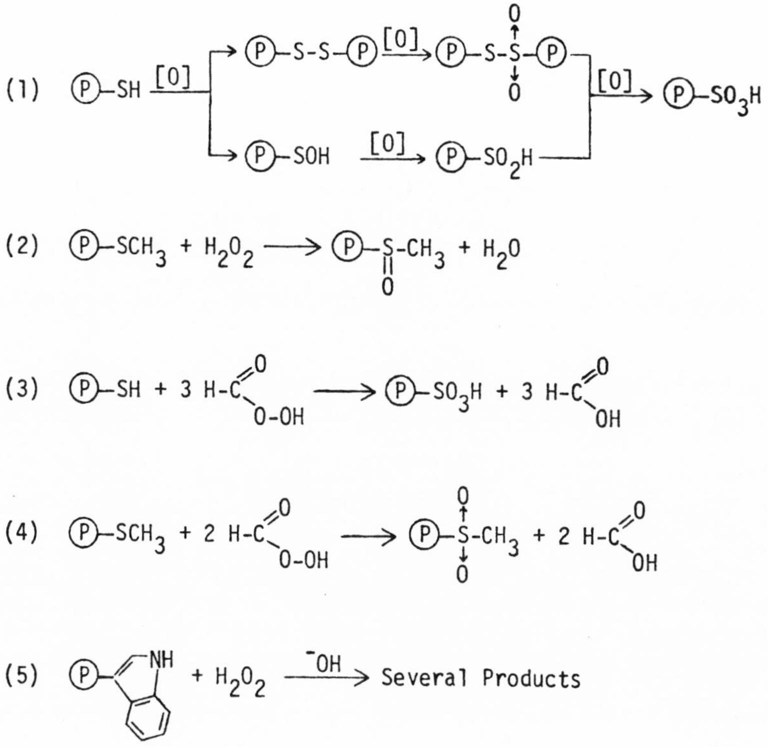

(1) (P)—SH [O] ⟶ ⎧ (P)—S–S—(P) [O]⟶ (P)—S–S—(P) ... [O]⟶ (P)—SO₃H
                  ⎩ (P)—SOH [O]⟶ (P)—SO₂H

(2) (P)—SCH₃ + H₂O₂ ⟶ (P)—S(O)—CH₃ + H₂O

(3) (P)—SH + 3 H–C(=O)(O–OH) ⟶ (P)—SO₃H + 3 H–C(=O)(OH)

(4) (P)—SCH₃ + 2 H–C(=O)(O–OH) ⟶ (P)—S(O)(O)—CH₃ + 2 H–C(=O)(OH)

(5) (P)—indole + H₂O₂ ⟶[⁻OH] Several Products

Holden-Day

Figure 5.   Oxidations of amino acids in proteins with peroxide (16)

inactivating the constituent inhibitors (19,20) (Tables IV and V).

Table IV. Protein Nutritive Value of Raw and
Cooked Red Gram (20)

Treatment	Protein Efficiency Ratio	Trypsin Inhibitor (units/100 mg)
Raw	0.68	10.8×10^{-3}
Cooked	1.43	Nil

Avi Publishing Company

Table V. Comparison of the Effects of Heating Methods on
Protein Nutritive Value of Soy Meal (20)

Treatment	Protein Efficiency Ratio	Available Lysine (%)
Unheated	0.63	58
Dry heat	1.00	53
Autoclave	1.75	46
Microwave	1.86	58

Avi Publishing Company

Deteriorative Reactions Involving Lysine

Amino groups are excellent nucleophiles, and there are many epsilon amino groups of lysines on proteins. Three of the most common types of deteriorations involving lysines are: non-enzymatic browning (Maillard) with reducing sugars, heat-induced damage involving isopeptide formation with the carboxyl groups of aspartic and glutamic acids or their amides (Figure 6), and formation of cross-linked products by interaction with alkaline degradation products, such as dehydroalanine.

The Maillard reaction involves attack of the nitrogen of the amino group on the carbon atom of the carbonyl, sometimes follow-ed by removal of water to produce the Schiff base (17) (Figure 7). Detailed coverage of the Maillard reaction is given else-where in this volume by Hodges (18), so only a few examples, particularly those with which the author has had some relation-

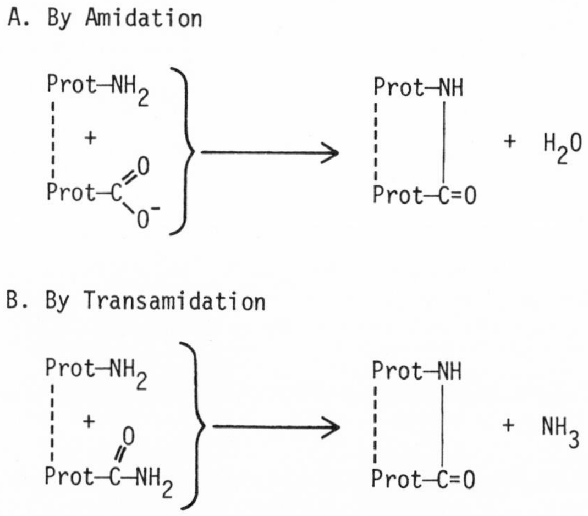

A. By Amidation

B. By Transamidation

Figure 6. Possible alternative reactions for formation of amide cross-linkages in proteins during heating

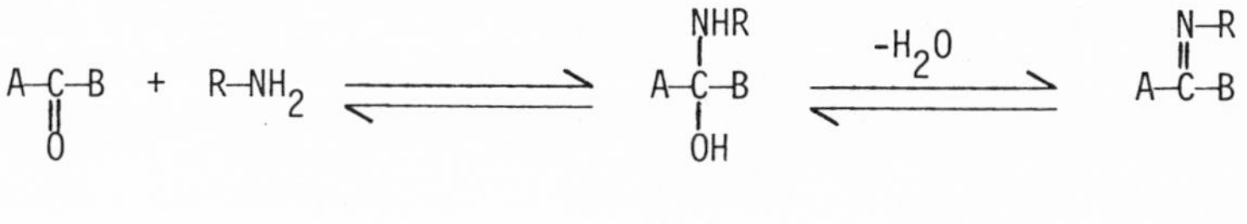

Advances in Protein Chemistry

Figure 7. Reaction mechanism of a strongly basic amine like an aliphatic amine or hydroxylamine with a carbonyl group (17)

ship, will be given here.

Maillard reactions are often considered to occur only under conditions where heat is applied or in dried samples stored for considerable periods of time. In some materials, however, such as liquid chicken egg white, where there is a high concentration of glucose (0.5%) and an alkaline pH (greater than 9), reaction of the amino groups of lysine with glucose occurs within a few days of storage of the intact shell egg at room temperature (21). These reactions result in changes in electrophoretic patterns (Figure 8), which caused confusion in genetic studies until the nature of the uncontrolled discrepancies was understood.

The products of at least one Maillard reaction caused the suffering of millions of non-scientists before the problem was unravelled. This was the reaction occurring in dried whole eggs, of which millions of pounds were eaten by American servicemen in World War II. When dried whole eggs were transported in the holds of ships to the South Pacific Islands and stored in jungle depots, the Maillard reaction usually resulted in products that were so extensively physically altered and had such vile and nauseating odors and flavors that many shipments had to be discarded. Much of this, consumed by agonizing army men who received a pile of such a disagreeable product as scrambled eggs in their mess kits, was frequently sufficiently bad to cause the men to vomit.

The author feels a close kinship to the dried egg development because he was on the receiving end of the devastating deteriorative reactions when the products were dropped in his mess kit for many months in New Guinea (now Papua and West Irian) and the Philippines in 1944-45, and because half-a-dozen years later he was nominally in charge of the research group at the U.S. Department of Agriculture's Western Regional Research Laboratory responsible for unravelling the cause. The research was led by Dr. Leo Kline. Before Dr. Kline's work the foul products had been attributed to a Maillard reaction involving the amino groups of the phospholipids and carbonyls formed by oxidations and hydrolyses of the lipids (23). As a result of these findings, dried eggs used by the military for the Korean war were acidified before drying and were packed with added sodium bicarbonate. The acidification slowed the Maillard reaction, and the bicarbonate served to neutralize the acid on reconstitution. The result was a more stable product, but some deterioration still occurred and the bicarbonate gave a soapy taste. Kline's group showed there was a much simpler explanation for the source of the carbonyls - the glucose (24). Glucose had been overlooked because the deteriorative reaction occurred in the lipid phase. Today the possibility of a reaction between the hydrophilic head of a phospholipid and a water soluble component seems so obvious as to be trivial, but thirty years previously it was not. The glucose, accounting for nearly 95% of the reducing sugar, could be removed by fermentation (25) or by oxidation with glucose oxidase (cata-

A 37°

B Control

C 37°

D Control

D 37°

E Control

F 37°

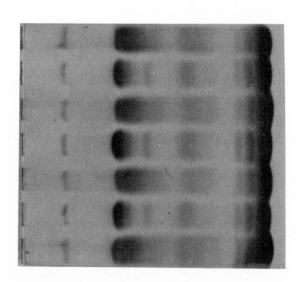

Journal of Biological Chemistry

Figure 8. Starch-gel electrophoretic patterns of incubated infertile eggs. Egg whites were all white Leghorn containing globulin A_1. Eggs were incubated at 37°C for 6 days or stored at 2°C for 6 days (controls). Letters refer to hen (22).

lase added to remove H_2O_2) to give a stable product when packed
in the absence of air (Figure 9).

There are, of course, many carbonyl compounds formed by hydrolytic or oxidative deteriorations of lipid constituents, and
most of these are potentially capable of entering into Maillard
reactions with proteins. One such product is reputedly malonaldehyde (26) (Figure 10).

Deteriorations Involving Disulfide Linkages

Sulfhydryl groups and disulfide bonds, and their inter-
relationships, are important groups affecting the properties of
the majority of proteins and are under continuous study by protein chemists. Indeed, the reduction of disulfides to form sulf-
hydryls, and the reoxidation of these to re-form the correct
pairings (Figure 11), are intimately related to the entire subject of protein conformation and conformational changes (27).
One of the long-enduring problems investigated in the author's
laboratory has been that of the deteriorative breakdown of thick
egg white and the egg white proteins on the surface of the yolk
membrane during the storage and/or incubation of shell eggs (10).
The breakdown can be simulated by the addition of mercaptans or
other disulfide-breaking agents (Figures 12 and 13).

Reduction and reoxidation have also been used to follow re-
activations of biologically active proteins. It was found that
an intermediate form of turkey ovomucoid (Figure 14), before complete oxidation, was actually slightly more active as an inhibitor of trypsin than was either the native protein or the completely reoxidized product.

Hydrolytic scissions of disulfides have been intensively
studied, particularly by the laboratory of Schöberl in Germany
(30). It has been shown that reactions such as these can occur
on the addition of small amounts of metal ions, such as copper or
mercury (31). Lysozyme, for example, is rapidly inactivated by
small amounts of cupric ion (Figure 15). But in many cases,
results of this nature have not been definitely shown to be due
to disulfide bond splitting. Other possible causes, such as
racemization, must also be considered.

Effects of Alkali on Proteins

Alkali has long been used on proteins for such processes as
the retting of wool and curing of collagen, but more recently it
has received interest from the food industry. Alkali can cause
many changes such as the hydrolysis of susceptible amide and
peptide bonds, racemization of amino acids, splitting of disulfide bonds, beta elimination, and formation of cross-linked
products such as lysinoalanine and lanthionine.

Our own laboratory has studied these reactions and, in particular, beta eliminations involving disulfides (Figure 16) and

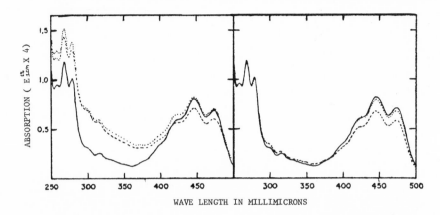

Food Technology (Chicago)

Figure 9. Effect of glucose removal on storage-induced changes in absorption spectra of ether extracts of stored dried eggs. The two samples illustrated were spray-dried powders stored 5 weeks at 37.5°C (25). Left side, untreated; right side, glucose-free. Control (——); air pack (— — —); N_2 pack (· · ·).

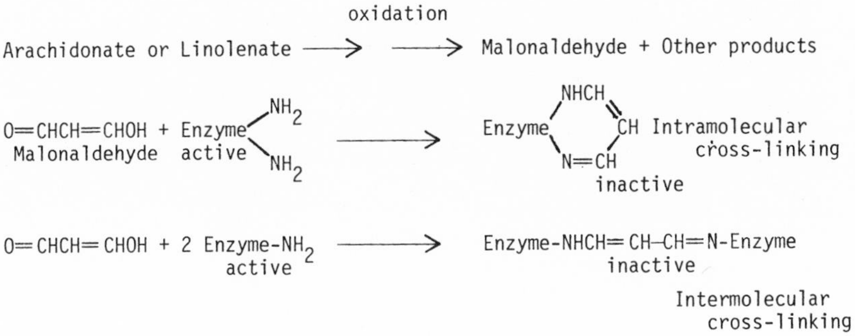

oxidation
Arachidonate or Linolenate ——>   ——> Malonaldehyde + Other products

$O\!\!=\!\!CHCH\!\!=\!\!CHOH$ + Enzyme active with NH_2 groups ——> Enzyme with NHCH / CH, N=CH Intramolecular cross-linking inactive
Malonaldehyde

$O\!\!=\!\!CHCH\!\!=\!\!CHOH$ + 2 Enzyme-NH_2 active ——> Enzyme-NHCH= CH–CH=N-Enzyme inactive

Intermolecular cross-linking

Biochemistry

Figure 10. The inactivation and cross-linking of RNase A by ethyl arachidonate and methyl linolenate (26)

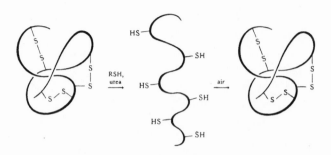

Holden-Day

Figure 11.   Schematic illustration of the fate of protein S-S bonds during reduction and unfolding followed by renaturation and reoxidation (16)

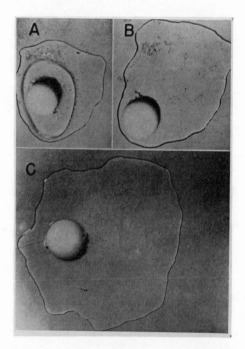

Poultry Science

Figure 12.   Effect of adding dilute thioglycol to broken-out eggs (top view). A, control after 4 hours at 25°C. B, egg with 30 ppm thioglycol shown after 4 hours at 25°C. C, egg with 300 ppm thioglycol shown after 4 hours at 25°C. Essentially the same results were obtained by a 2-hour treatment with the same amounts of reducing agent (28).

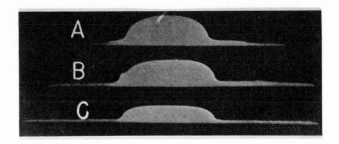

Poultry Science

Figure 13.   The same conditions as in Figure 12, showing side view (28)

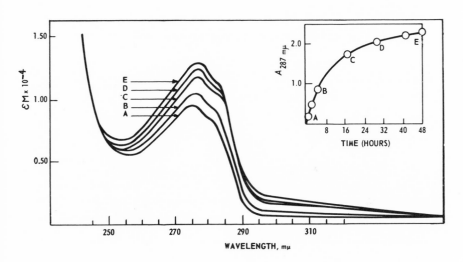

Biochimica et Biophysica Acta

Figure 14.   Changes in ultraviolet absorption spectra of turkey ovomucoid after reduction and after various periods of reoxidation. Protein (0.7 mg/mL, containing 8–10% water) was dissolved in 0.006M Tris buffer adjusted to pH 8.3. Incubation was at room temperature for the following times (hours): A, zero (starting); B, 4; C, 17; D, 29; E, 48 (29).

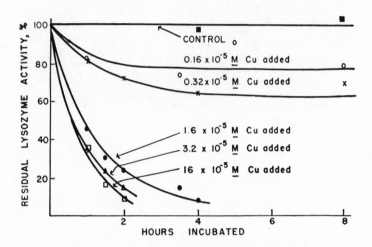

Archives of Biochemistry and Biophysics

Figure 15.   The influence of copper concentration on the inactivation of lysozyme. The lysozyme concentration was 3.3×10^{-6}M; the temperature was 47°C; and the buffer was borate adjusted to give pH 9.3 at 47°C (31).

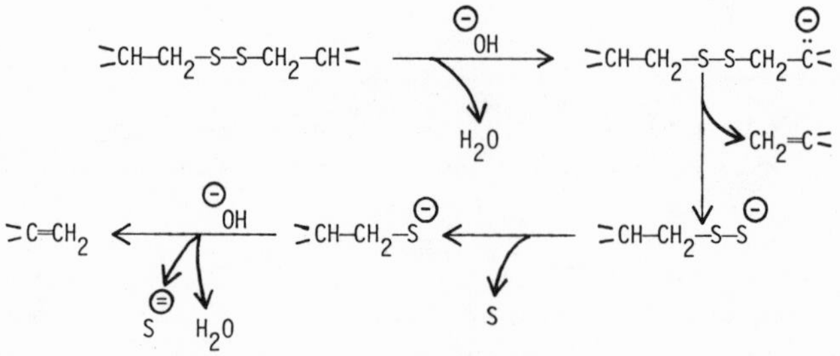

Figure 16.   β-Elimination scheme for disulfides on alkali treatment

the formation of deteriorative products (32) (Table VI). Alkali treatment is sufficiently important that a separate chapter is devoted to it in this volume (33).

Table VI. Analysis of Alkali Treated Lysozyme[a]

Amino Acids Lost (mole/mole)		Products Formed (mole/mole)				
Cystine	Lysine	Lanthionine	Lysino-alanine	β-Amino-alanine	S^{-2}	S
4	5	1	5	1	3	4

[a]Data from reference 32. Conditions were 0.1 $\underline{M}$ NaOH, 50°C, 24 hours.

Journal of Agriculture and Food Chemistry

Photooxidation Reactions

Photooxidations are not normally considered a protein deteriorative reaction because they usually go unseen or are found only when purposely contrived, such as in the chemical modification of proteins (Figure 17). However, the possibility of their occurrence in foods, particularly those containing added dyes, should not be overlooked. Several of the important amino acid side chains are readily modified, including the sulfhydryl, imidazole, phenoxy, indole, and thio ether (Figure 18). More general and detailed coverage is provided in another article in this volume (34).

Our own laboratory has studied the photooxidation of the imidazoles of histidines in the homologous proteins, human serum transferrin and chicken egg white ovotransferrin. These proteins contain two separate sites for binding iron, each of which apparently involves two histidines, three tyrosines, and perhaps a bicarbonate bonded between the iron and an arginine (35) (Figure 19). When the iron-free protein was photooxidized with visible light and methylene blue as the activating dye, the iron-binding activity was rapidly lost, apparently with the destruction of histidines in the iron-binding sites (Table VII). The ratio of the rate of loss of histidine to the rate of loss of iron-binding capacity was found to be 2.2, suggesting that there are two histidines in the iron-binding site. The pH profile for inactivation also indicated that the destruction of histidine was responsible for the loss in activity. Since the addition of 10 mM NaN$_3$ prevented the inactivation, singlet oxygen is apparently involved in the photooxidative inactivation. These results from photooxidation indicating the essentiality of histidines were confirmed in chemical modifications using ethoxyformic anhydride, a reagent with reasonable specificity for histidines (16).

$$DYE \xrightarrow{\text{h}\nu} DYE^*$$

$$DYE^* + O_2 \longrightarrow DYE + O_2^*$$

$$O_2^* + \text{AMINO ACID} \longrightarrow \text{OXIDIZED PRODUCTS}$$

Figure 17.   Dye-catalyzed photooxidation of amino acid side chains in proteins

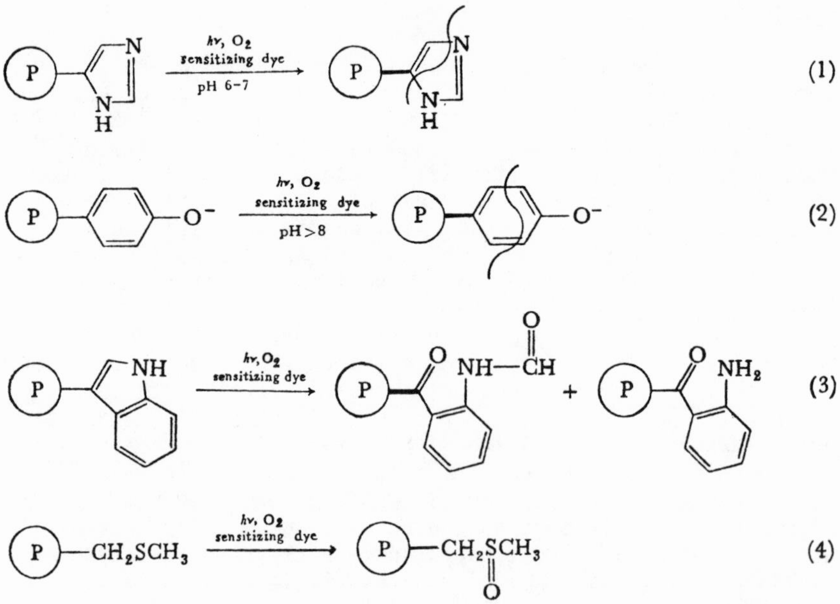

Holden-Day

Figure 18.   Photooxidation pathways for histidine (1), tyrosine (2), tryptophan (3), and methionine (4) (16). Wavy lines through structures indicate ring scissions. Cysteine is oxidized to cystine, in some cases, without sensitizing dye; cystine may also be oxidized.

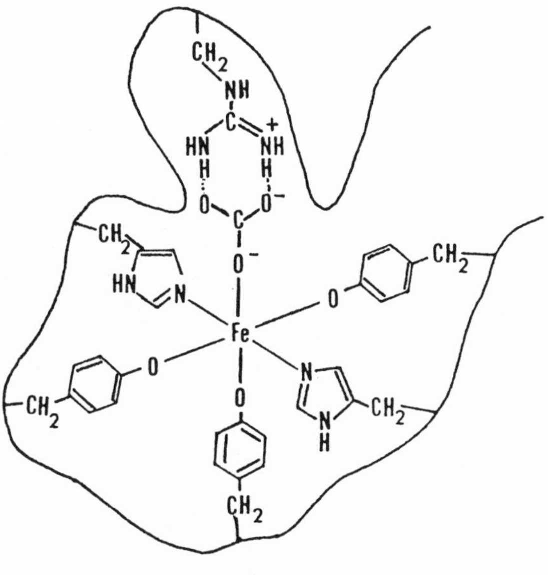

Figure 19. A model for the anion- and iron-binding sites of transferrin depicted assuming an interlocking-site hypothesis. The protein furnishes five ligands to the metal in the iron binding site; three tyrosines and two histidines. The carbonate ion binds to an arginine in the anion-binding site and functions as a sixth ligand to the metal center. The carbonate forms a bridge between the metal- and the anion-binding sites in the active center (36).

Table VII. Loss of Activity of Ovotransferrin on Photooxidation[a]

Time (minutes)	Loss in Activity (%)	
	No azide	Azide
20	16	—
40	31	< 5
100	58	< 5

Biochemistry

[a]Data from reference 35.

Undesirable Chemical Products Formed in the Processing of Foods and Feeds

There have been a number of well-documented examples of the formation of chemical derivatives of the amino acid side chains in proteins during the processing of foods and feeds. It would be wrong to believe that there are not many more examples of at least minor amounts of undesirable substances introduced by similar procedures, but, fortunately, none of major importance have surfaced other than the two discussed below.

The first of these well-documented examples was the formation of a toxic product, methionine sulfoxamine. This was formed on the bleaching of flour with nitrogen trichloride, popularly known as agene (37) (Table VIII). Agene was used until the early 1940's for the bleaching of flour to produce the white flour desired by American consumers. When agenized flour was fed at high levels to dogs, the dogs developed the central nervous disorder popularly called "running fits". In this disorder, the dogs were excitable and, on minor external stimuli, would throw themselves about in a way characterized by the name. When these observations were made known, the use of agene for bleaching flour was immediately discontinued. There were no experiments on humans, of course, and whether or not any damage to humans was ever observed remains unknown.

The second major example of a process-induced chemical side reaction was the formation of dichlorovinylcysteine. Dichlorovinylcysteine was formed when soybean oil meal was extracted with trichloroethylene for the removal of the fat in the production of animal feedstuffs (39) (Table VIII). The symptoms were first observed in cattle fed the extracted meal. The cattle developed hemorrhagic symptoms and many died. This process was discontinued. Although the trichloroethylene-processed soybean meal was entirely designated for animal feed, there was the possibility that small amounts could have found their way into human

Table VIII. Toxic Compounds Produced in Foods and Feeds by Chemical Modifications (38)

Chemical		Use	Product	
Nitrogen trichloride (Agene)	NCl_3	Bleaching flour	Methionine sulfoximine	$\begin{array}{l} -C=O \\ -C-CH_2-CH_2-S-CH_3 \\ -NH \end{array}$ (with O and NH on S)
Trichloro-ethylene	$ClCH=CCl_2$	Extraction of oil from soybean	Dichloro-vinyl-cysteine	$\begin{array}{l} -C=O \\ -C-CH_2-S-CH=CCl_2 \\ -NH \end{array}$

Avi Publishing Company

food, but there are no records of this, and there are no records
of any human ailment ascribed to this agent.

There is always a possibility of at least minor changes, important or unimportant, creeping into processes when chemicals
are used. In the enzymatic oxidation of glucose to gluconic acid
to remove the carbonyl groups of glucose in certain foodstuffs,
such as in the preparation of dried egg white, hydrogen peroxide
is a product of the reaction, requiring the addition of catalase
for its decomposition. Hydrogen peroxide (see Figure 5) is occasionally used as a sterilizing agent and is even added to foodstuffs.

Active Site Selective Reagents - Naturally Occurring Toxins and Laboratory Tools

The term "active site selective reagents" is used to describe several different kinds of reagents that react covalently
in the active center of an enzyme. The term is widely used to
include agents that react in a particular part of a protein doing
a specific task in some kind of biochemical process. With an
enzyme the reagent usually resembles a substrate and by some
reaction remains in the active center, thereby inactivating the
enzyme or leaving a piece of the reagent in the center. This
procedure is used to label or find the groups that are in the
active center as well as to inactivate the enzyme, although it is
not necessary that inactivation occur.

There are many different definitions describing the active
center, active site, combining sites, and allosteric sites of
enzymes (16). One of the more commonly used definitions is that
the active center of the enzyme is that area or that place in the
enzyme which contains the active site of an enzyme and everything
else that is in that area, usually meaning at least part of the
binding site for the substrate and other groups that are there in
order to maintain structure, react with water, or provide a hydrophobic pocket, etc. The active site in turn is usually taken as
that part of the enzyme which does the work, i.e., the catalytic
process. The term "active site selective reagents" therefore
really should be "active area selective reagents," but the former
term is so extensively used that we will continue to employ it
here.

Active site selective reagents can be classified in various
manners. One way is to divide them according to how they react
(Table IX). In such a classification one finds substrates that
can be covalently attached by chemical treatment of the enzyme
while it is catalyzing some change in the substrate. An example
is the reaction of functional amino groups by the enzyme muscle
aldolase acting on glyceraldehyde and reduced by cyanoboro-
hydride (16).

A second type of active site selective reagent is when there
is a pseudosubstrate, such as diisopropylfluorophosphate. Acting

Table IX. Active Site Selective Reagents

Type	Mechanism
Substrate	Normal intermediate product can be covalently attached, e.g., by reduction
Pseudosubstrate	Product is poor leaving group, e.g., DFP
Affinity Reagent: General	"Double-headed" - one like substrate, other chemically reactive
Photoaffinity	"Double-headed" - one like substrate, other converted to chemically reactive group by photoactivation
Product of Enzyme Reaction-"Suicide" Reagent	Is a substrate, part of which is converted to chemically reactive group by enzyme catalysis

on the enzyme trypsin, such a pseudosubstrate may contain a poor
leaving group and thereby remain on the active site as a diiso-
propylphosphoryl ester of the serine in trypsin.

A third type is the one that includes a number of subclassi-
fications, all of which come under the general term "affinity
substrates (reagents)". The two types described above also have
affinity characteristics, but these latter ones are different. A
sketch of an affinity reagent is shown in Figure 20. In this,
the binding group is what the enzyme recognizes and binds, while
the group marked X, the covalently reactive group, now is able to
form a covalent bond somewhere in the vicinity of the active
center or at its periphery, providing, of course, that there is a
suitably susceptible amino acid side chain in these locations.
The affinity reagent is therefore always a double-headed one, one
head resembling the substrate and the other head the working head
to form a covalent bond. There are at least three kinds of
affinity reagents: those in which the covalently reacting group
is already present in the reagent, those in which the covalent
reactive group must be generated by an external action such as
photoactivation, and those in which the enzyme itself generates
the reactive group. These latter have been termed "k_{cat} reagents" because their interaction occurs as a result of the
enzymatic catalysis to form the reactive group, or "suicide reagent" because the enzyme kills itself by a catalytic action
(41-44). These have a very much higher specificity than other
affinity reagents because they not only have the specificity of
binding in common with the others, but they also have the specificity of catalysis which the others do not have. In this
respect they should thus be a "perfect drug".

These "suicide reagents" will be described in detail else-
where in this volume (44), so only one phase will be briefly
mentioned here. Of particular interest to food and nutrition re-
searchers are the naturally occurring toxins which involve a
"suicide" mechanism (42). Some of these can be consumed in foods
or feeds and commonly occur in a number of different plant
sources. A very common toxin is the beta-aminopropionitrile
present in lathyritic legumes, and another is the wildfire toxin
(42) (Figure 21).

Chemical Deteriorations to Purposely Derivatize Proteins

Under this heading might be placed many of the reactions already discussed, but there are several that fit more appropriate-
ly in such a classification. One of these, with which the author
has been associated, is the formation of inactive derivatives of
proteolytic enzymes by alkaline beta elimination of a derivative
of the active site serine of trypsin (45) (Figure 22). This
modification uses an affinity reagent followed by a second chemical modification, the alkaline beta elimination, to form the
product. The products of the reaction with trypsin and chymo-

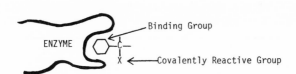

Figure 20. Diagram of principle of affinity labelling of a reactive site. In affinity labelling there is (a) a binding group that resembles the type of substance (substrate, antigen, etc.) with which the protein normally interacts specifically, and (b) an additional group, a covalently reactive group, capable of forming a covalent bond in the reactive site. Affinity reagents are usually classified into three different types: general affinity, photoaffinity, and "suicide" affinity (40).

Accounts of Chemical Research

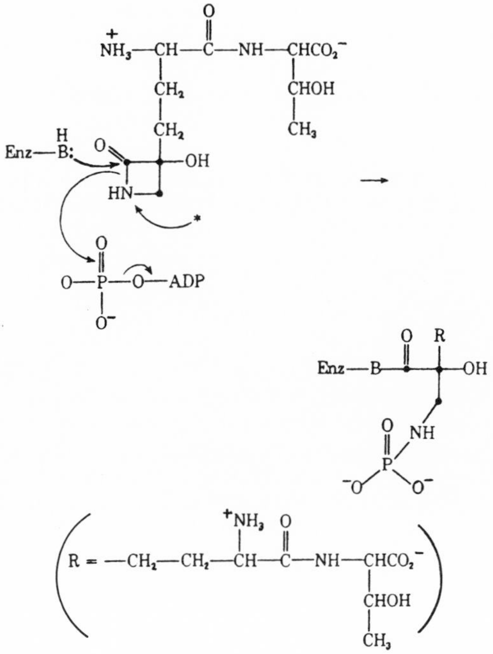

Figure 21. Inhibition of glutamine synthetase by wildfire toxin (42)

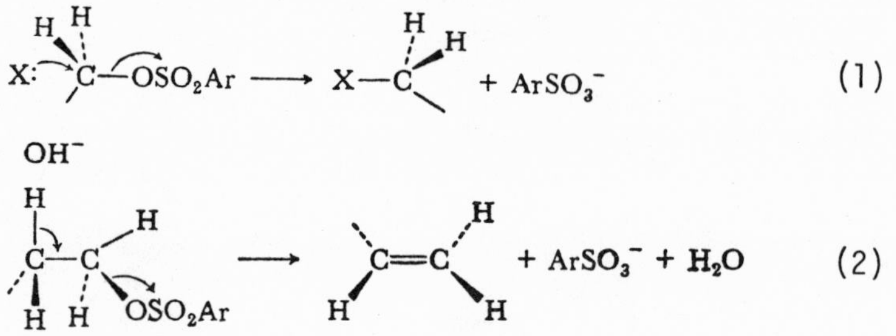

$$X:\overset{H}{\underset{H}{C}}\!\!-\!\!OSO_2Ar \longrightarrow X\!-\!C\overset{H}{\underset{}{}}\!\!\overset{H}{} + ArSO_3^- \qquad (1)$$

$$\underset{\overset{|}{C}}{OH^-}$$

$$\underset{H\ \ H}{\overset{H}{C}}\!-\!\overset{H}{C}\!-\!OSO_2Ar \longrightarrow \overset{}{C}\!=\!\overset{H}{C}\!\! + ArSO_3^- + H_2O \qquad (2)$$

Figure 22.   *Displacement of an aromatic sulfonate (weakly basic) by (1) nucleo-philic attack or (2) β elimination with alkali (16)*

trypsin are called anhydrotrypsin and anhydrochymotrypsin, respectively. The reaction is similar in mechanism to the alkaline
beta elimination of O-phosphoryl or O-glycosyl groups described
above. The products, anhydrotrypsin or anhydrochymotrypsin, are
very useful in enzyme chemistry because the overall structure and
conformation of an enzyme is very little affected. They have
been used in several different studies, one of which is the
interaction of proteolytic enzymes with specific protein inhibitors. The anhydro derivatives will form highly associated complexes with the inhibitors in a manner very similar to that of
the native catalytically active enzymes (Table X). In fact, in
some cases they may be as effective, or even better, in combining
with the inhibitors than the native enzyme. These data have been
used as evidence that catalytic action, including formation of a
tetrahedral adduct or an enzyme acyl bond, is not necessary for
the formation of the inhibitory complex (47).

   Another similar type of reaction has been the use of an
affinity reagent (2',3'-epoxypropyl β-glycoside of di-(N-acetyl-
D-glucosamine) to react with a carboxyl group of an aspartic acid
in the active center of the enzyme lysozyme (48) (Figure 23).
Then this reagent can be removed from the enzyme by reduction.
Since the bond between the affinity reagent and the carboxyl
group of lysozyme is an ester bond, the carboxyl group of aspartic acid of the enzyme is now changed by reduction to an
alcohol. The new residue is therefore an aspartic acid with a
carboxyl group changed to a hydroxyl to give 2-amino-4-hydroxy-
butyric acid (homoserine). The homoserine lysozyme has properties so similar to those of the original enzyme that it forms
tight complexes with substrates (Table XI).

   A modification embodying several of the different deteriorative reactions discussed in this article was recently studied in
our laboratory (50). Two different avian ovomucoids with different inhibitory properties against proteolytic enzymes were modified by the alkaline beta-elimination reaction so as to form new
covalent cross-links consisting of lanthionine and lysinoalanine.
One ovomucoid was turkey, which has a double-headed character
with independent sites for forming an inhibitory complex with
bovine trypsin at one site and bovine alpha-chymotrypsin at the
other site. Both of these sites are relatively strong binding
sites as compared to the strengths of binding of other inhibitors. In addition, the alpha-chymotrypsin binding site will also
accept the bacterial enzyme subtilisin, which has an affinity for
the inhibitor of about the same order of magnitude as does alpha-
chymotrypsin. Consequently, the two enzymes compete about
equally for the same site. In contrast, penguin ovomucoid has
the same two sites as turkey ovomucoid, one for trypsin and one
for chymotrypsin, but the relative affinity for the different
enzymes is quite different. The trypsin site is relatively weak,
and the chymotrypsin site is quite strong for subtilisin but much
weaker for alpha-chymotrypsin. When penguin and turkey ovomu-

Table X. Comparison of Association Equilibrium Constants
for Inactive and Active Enzymes (46)

Inactive Enzyme	Inhibitor	$K_a$inactive/ $K_a$active	$\Delta G_{inactive}$ - ΔG_{active} (kcal)
Anhydro-trypsin	Bovine pancreatic (Kunitz, BPTI)	>0.2	<0.9
	Reduced BPTI	2.0	-0.4
	Bovine pancreatic (Kazal)	0.097	1.4
	Chicken ovomucoid	0.025	2.2
	Soybean inhibitor	0.010	2.7
	Lima bean inhibitor (unfractionated)	0.008	2.9
Anhydro-chymotrypsin	Potato inhibitor	1.0	0.0
	Lima bean (III)	>200.	<-3.0
	Bovine pancreatic (Kunitz)	1.0	0.0
Methyl-chymotrypsin	Turkey ovomucoid	0.010	2.7
	Duck ovomucoid	0.014	2.5
	Golden pheasant ovomucoid	0.010	2.7

Springer–Verlag

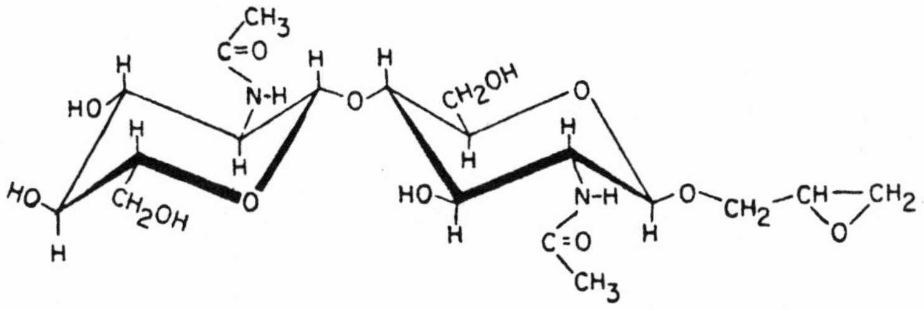

Journal of Biological Chemistry

Figure 23. Structure of 2′,3′-epoxypropyl β-glycoside of di-(N-acetyl-D-glycosa-mine) (48)

Table XI. Enzymatic Activities and Association Constants of
Lysozyme and Its Derivatives (49)

Lysozyme	Relative Enzymatic Activity[a]	Association Constants (M^{-1}), pH 6.7, 23°C[a]	
		$(GlcNAc)_3$[b]	$(GlcNac-MurNAc)_2$[c]
Native	1.00	1.1×10^5	4.2×10^3
Regenerated[d]	0.97	3.3×10^3	1.4×10^3
Hse^{52}[e]	0.10	1.2×10^4	6.3×10^3

[a]The experimental error of the values given in the table is ±
10%.

[b]GlcNAc, N-acetyl-D-glucosamine.

[c]MurNAc, N-acetylmuramic acid.

[d]Regenerated from the ester by weak alkali.

[e]Lysozyme in which aspartic acid-52 was replaced with homoserine
by reduction of the ester.   Proceedings National Academy of Science

coids were treated with alkali or thiocyanate to form a new
covalent cross-link, products were obtained which still retained
their biochemical affinity for one or more of the enzymes, but
with important differences (Figure 24, Table XII).

These results also serve as an example of difficulties that
can be encountered in analyzing for deteriorations by determining
losses of biochemical activity of proteins on exposure to alkali
or merely to higher pH and moderate heat. Proteins like the
inhibitors of proteolytic enzymes, which have activities against
several different enzymes, might then show large differences in
loss of activity depending upon which enzyme and perhaps even
which assay condition was employed for the determination of their
activities. Assays against one enzyme could show no losses in
activities, leading to a possible deduction that there were no
effects on the protein, while use in the assay of another enzyme
could show as much as complete destruction.

Reversal of Protein Deteriorations

Although reversibility of the deteriorative reactions of
proteins is usually not a practical possibility, particularly
with mixtures of proteins and proteins in complex biological

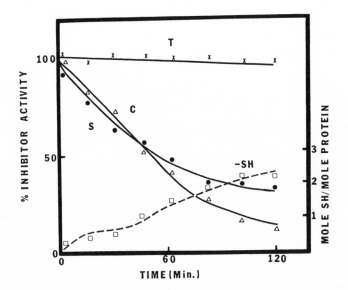

Figure 24. The effect of disulfide bond modification of turkey ovomucoid by alkali on inhibitory activity against trypsin (T), α-chymotrypsin (C), and subtilisin (S). Turkey ovomucoid (0.10 mM) was treated with alkali (100 mM NaOH) at 23°C. Sulfhydryl content (moles per mole of protein) (−SH) is shown (− − −) (50).

Table XII. Loss and Formation of Amino Acids and Thiocyanate
in Cyanolysis-Treated Penguin Ovomucoid (50)

Time[a] (min)	Loss of cystine[b]	Formation of lanthionine[b]			Formation of iminothiazolidine[c]	Formation of thiocyanate[d]
		meso	d,l	Total		
30	0.04	0	0	0	0	0.45
400	2.2	0.34	0.62	0.96	0	2.5
600	4.7	0.45	2.1	2.55	0	4.1
1400	5.5	1.2	4.9	6.1	0	5.3

[a]Time of cyanolysis of penguin ovomucoid.

[b]Moles of amino acid lost or formed per mole of protein.

[c]Cyclization product of β-thiocyanoalanine.

[d]Free thiocyanate anion formed (mole/mole protein).

International Journal of Peptide and Protein Research

systems, in some cases reversal can be achieved. By the very
nature of the process of denaturation and renaturation, renaturation is the one reversal that should be the easiest to achieve.
In pure systems, and with many proteins, this is now frequently
and easily done. Descriptions of this have been given in the
previous section on denaturation. However, when the denaturation
is accompanied or followed by chemical reactions, complexing with
other substances, or even insolubilization, reversibility of the
denatured protein may be impractical.

The reversibility of the reductive scission of disulfide
linkages has also been discussed in a previous section. Such a
reversal is intimately related to the renaturation process because the thiol groups must accept their correct original part-
ners in the formation of the disulfide linkage. In mixtures of
different proteins or with other substances, the correct pairing
may be prevented.

Reversibility of the chemical modifications of proteins is
frequently a desired objective of the scientist using chemical
modification as a tool (16,51). The objective here is to chemically change the protein so as to affect its physical or biochemical properties and then to reverse this change so as to
obtain the original protein again. In chemical deteriorations
which occur inadvertently or "naturally", however, reversibility
is usually hard to attain. An example of one that is of a practical nature is the oxidation of the thio ether of methionine to
the sulfoxide (Figure 5). This type of reaction could be encountered in the use of hydrogen peroxide as a sterilizing agent.
This reversibility is easily achieved by treatment with low
concentrations of a reducing agent.

Detection and Determination of Deteriorated Proteins

The detection and determination of deteriorated proteins are
two areas that plague the basic researcher, the clinician, and
the practical food technologist. Simple, workable methods are
usually just not available. Of all the questions put to the
author by members of the food industry, those on this area head
the list in frequency.

The detection and determination of deteriorated proteins is
probably one of the most difficult problems in protein chemistry
today. The main reason is that it is usually easier to determine
native properties, particularly when those properties include a
biochemical function, than it is to determine a deteriorated and
inactive protein. In addition, there are almost always no stand-
ards to use as a measuring stick for the deteriorated protein.
When the deteriorative reactions are extensive and more than a
fraction (30-40%) of the protein undergoes deterioration, the deterioration can frequently be monitored by determining the amount
of native protein remaining, providing, of course, that the deteriorated protein does not interfere with the determination of

the native one. However, since many deteriorative reactions may
involve only a minor fraction of the total protein, or even of an
individual protein, it is usually impossible to determine the
amount of protein deterioration by determining the amount of unmodified protein. Furthermore, deteriorative reactions frequently involve such a minor change in the physical and chemical properties of the protein that suitable methods are not available to
easily monitor these minor changes.
    In spite of these many limitations, there are a number of
approaches that have been successfully used for certain deteriorative reactions (Table XIII). The most obvious approach is to
determine the biochemical or biological activity of a protein
when the protein has such a property. The difficulty in these
determinations is that the deteriorated protein may have a biochemical activity that is changed or attenuated without loss of
activity. In such a hypothetically modified protein, a biochemical determination might show that there is only 50% activity
present. However, such a determination could not differentiate
between complete inactivation of 50% of the molecules or a 50%
reduction in activity of 100% of the molecules by an attenuation
rather than an inactivation. A different test would be necessary
in order to "count" the number of biochemically active molecules.
For this purpose, the enzyme chemist sometimes uses an active
site titrant, which is a type of an affinity reagent (16).
    Probably the oldest physical method used for changes in pro-
teins is the measurement of solubility. Deteriorated proteins
have traditionally been thought to show decreases in solubility,
although frequently increased solubility may also occur, such as
is usually found with proteolysis. Along with solubility changes
go changes in viscosity, a physical change that, in certain
systems, is easily determined. With more purified systems,
ultraviolet absorption and optical rotation (circular dichroism)
are valuable procedures. Ultraviolet absorption is a simple
test, and changes in absorption are a reflection of changes in
protein conformation or, in some cases, the formation or destruction of chromogens.
    A more general method that is applicable for many changes in
protein properties is that of electrophoresis, using the more
modern and simplified gel electrophoretic techniques (52,53).
Micro gel electrophoresis can be used to show changes in a charge
on a protein, changes in conformation, and the scission of peptide bonds. Direct gel electrophoresis will show charge changes,
as used in studies on the Maillard reaction in which the amino
groups of proteins lose their charge or have charge changes
on formation of the reaction products with glucose (Figure 8).
Electrophoresis with denaturing additives and disulfide bond
scission agents, such as thiols, is useful in testing for the
splitting of peptide bonds. Aggregations and cross-linkings can
be shown in sodium dodecylsulfate gels with or without addition
of thiol compounds, depending upon the purpose of the experiment.

Table XIII. Simplified Outline for Detections of Some Protein
Deteriorations

Deteriorations	Effects	Methods
Denaturation	Solubility changes, release of lipid, etc., racemization	Physical-either simple, like solubility, or more sophisticated, like CD
Chemical	Hydrolysis of peptides or amides, release of sugars, etc.	Direct physical or chemical analysis, e.g., titrations; free substances
	Formation of internal isopeptides	Titrations and peptide mapping
	β Eliminations and cross-linking	Direct spectrophotometry, volatile and amino acid analysis
	Oxidations	Direct spectrophotometry, chemical modification, amino acid analysis, peptide mapping
	Interactions with substances, e.g., sugars	Spectrophotometry, electrophoresis, amino acid analysis

The power of these relatively simple techniques is seen in the determination of equilibria between denatured and native states by electrophoresis in different concentrations of urea (54). Since unfolding of a protein is usually accompanied by a decrease in mobility through a gel, conformational properties can be obtained as a continuous function of urea concentration. Other refinements using more sophisticated detection procedures involve immunoelectrophoresis and procedures for the development and detection of enzymes in gels with enzyme staining procedures.

Because thiol groups of cysteine are easily determined by micro methods and the reactivity of the thiol groups of proteins is frequently affected by the protein's conformation, the determination of sulfhydryl groups has been used extensively as a detection procedure for changes in protein conformation. In complex systems (e.g., a food product), however, such methods are often useless.

Changes in optical rotation of the individual amino acids as a result of racemization can now be determined by several micro methods (55). More advances in these procedures will undoubtedly be forthcoming soon.

Sometimes an investigation of the deteriorative reaction has necessitated the development of new methods, or at least the refinement of older methods, in order to quantitate the deteriorations. This was found necessary in studies in the author's laboratory on the treatment of proteins with alkali (56). In these studies, the quantitation of sulfur balances showed that not all the sulfur was accounted for in the form of organic products. A probable product was H_2S, with the possibility of some elemental sulfur also being formed. The determination of these two substances in the small concentrations in which they would be found in proteins used in test tube experiments was difficult by the methods available at that time. Adaptations of existing methods for hydrogen sulfide, combining the distillation-purging procedures used for inorganic studies with the microanalytical procedures for thiols and proteins employing DTNB (5,5'-dithiobis-(2-nitrobenzoic acid)), were used to make a satisfactory method (56). Very low concentrations of hydrogen sulfide (0.14 µg/ml; 4.4 X 10^{-6} $\underline{M}$) could easily be determined with DTNB. In the presence of proteins containing different amounts of sulfhydryl groups reacting with DTNB, approximately 80 to 95% of the hydrogen sulfide could be recovered. The determination of sulfur was based on its high solubility in hydrocarbons and its extinction in the near ultraviolet (Figure 25). Its high solubility in n-heptane, together with its very low solubility in water and its absence of a charge, allow for easy removal of interfering colored agents. Amounts as low as 0.2 µg/ml are easily determined in the presence of proteins. This has been one example of the need to delay a research project in order to develop or adapt methods for studying protein deterioration.

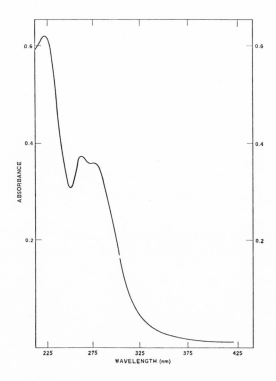

Analytical Biochemistry

Figure 25.   Quantitation of sulfur. Sulfur was extracted with n-heptane from the reaction mixture, and the heptane was washed twice by extraction with phosphate buffer. The absorbance of sulfur (14.3 μg/mL) in n-heptane was plotted as a function of wavelength. The absorption at 225 nm was used for calculation of the amount of sulfur present (56).

The fractionation and purification of deteriorated proteins
is undoubtedly one of the least successful techniques. This is
simply because all of the methods that have been developed, with
very few exceptions, are directed toward purifying the unde-
teriorated protein. The methods available are usually based on
some particular biochemical activity of the protein, usually enzyme activity, and sometimes an affinity column or affinity adsorbent could be used to separate the native protein from the
deteriorated one. Quite often a good affinity adsorbent is
unavailable. This procedure, however, does not always work
properly even when an adsorbent is available, because the deteriorated protein may possess some activity or an affinity for
the adsorbent even though it has lost its natural enzyme activity
(see Figure 24). The antigen-antibody reaction can also be used
by means of precipitation with antibodies against the native
proteins or adsorption on the immobilized antibodies. But here
again, the specific antibody must be available, and the deteriorated protein may retain so much affinity for the antibody that
differential separations will be impractical in some cases.

Conclusions

    There are many possible conclusions to draw from an examina-
tion and summarization of what is known about protein deteriorations. Two main conclusions are:
1)   Scientists are presently using methods which discriminate
against deteriorated proteins. Almost all the techniques developed are aimed at purifying native proteins or getting rid of
deteriorated ones. In current purification procedures, deteriorated proteins are usually eliminated or, at best, are separated
only in small quantities and as mixtures of more than one deteriorated product.
2)   Deteriorations of proteins should receive greater attention
in the future. Medical scientists are becoming increasingly
aware of the importance of deteriorated, or at least modified,
proteins in aging processes (such as connective tissue) (57,58)
and in other changes related to protein synthesis and tissue development (59). Protein deteriorations should find more use in
other areas, such as the estimation of the age of bones and eyes
and in diseases of the eye, as shown in the novel findings of the
laboratory of J. L. Bada on the racemization of amino acids
(60,61). These deteriorations have even been used for the estimation of geological age (62).

Some Particular Areas for Future Investigations on Protein
Deteriorations

    The subject of protein deteriorations encompasses most of
protein chemistry and is a wide-ranging and difficult topic. It
is hoped that from this article and the other articles in this

monograph it is evident that the importance of deteriorative reactions is great, and future research will be an ambitious undertaking.

The following are some possible areas for future investigations of protein deteriorations:

A.   Living systems

1) Biochemical control systems (i.e., zymogens, activators, etc.). It is in this area that large advances should be forthcoming in the next decade or so. The sophisticated selective breakdowns of proteins to form functional products are among the most important biochemical control reactions found in tissues.

2) Aging. The biochemistry of senescence is a very immature subject at this time. The majority of the changes that occur are not known, including the reason for the time clocks, but some advances have been made (59). Intimately related to the aging process are the related processes of disease.

3) Environmentally induced deteriorations. Currently most emphasis in studies of environmental effects on biological systems is on toxicity and the onset of malignancy. Malignancy is usually directly associated with nucleic acids, but, at least in some cases, nuclear and other proteins may be involved as well. This is an area requiring intensive research. Toxicities usually involve the interaction of the toxic agents with some protein system, particularly enzyme systems.

4) Pharmaceutical effects. Most pharmaceutical agents involve drugs which act either at cell receptor sites or directly on enzymes. In both cases proteins are involved. With many drugs, there is a man-directed deteriorative reaction to inactivate a particular target (43). Usually, the greater the specificity of inactivation, the more specific the drug.

5) Tissue and organ preservation. The preservation of tissues and organs for future use is becoming an important part of the armament of medical technology (63). With most procedures, low temperatures, or even freezing procedures, are required. Many physical and chemical changes occur during the freezing and thawing of tissues (64), most of which are unknown or, at best, are handled by empirical procedures (such as the addition of stabilizing substances). There is a need here for extensive, fundamental researches on low temperature preservation of tissues and organs, as well as the development of milieu for their storage and preservation at higher temperatures.

B.   Foods and feeds

The post-harvest or post-slaughter changes that occur in plant and animal materials are extensive, but they are poorly understood. This comparative lack of information necessitates a technology of processing and storage which is extensive, cumbersome, and frequently empirical. Food and feed material must be preserved without microbial spoilage, and the physical changes in the products that result in unwholesome or undesirable characteristics must be eliminated. Many of the procedures used involve drying, but this uses vast expenditures of energy and can cause extensive chemical changes in proteins as well as extensive denaturation (65), as with dairy or poultry products such as dried eggs. The global involvement of World War II resulted in an entirely new research effort in the United States - a very extensive program under the Army Quartermaster Corps for improvements in the preservation, storage, and transport of foods, which frequently involved very harsh environmental conditions.

Most of this type of research has now been taken over by the fast food industry. There is a great additional need for fundamental studies in these areas to develop processes based on sound principles rather than the empirical measures necessary during World War II, and to produce stable foods for feeding the world's millions rather than just the affluent purchasers of fast foods.

The author discussed the matter of research support in these areas two decades ago; the type and amount of financial support does not appear to have changed much since that time (66). Fundamental research on the preservation of food is supported to only a limited degree by research grants. Most of the high quality research has been indirectly supported by grants from the National Institutes of Health, frequently on disease-related projects. This is a shameful reflection of priorities on the value of basic food research.

C.   Methods for detection, characterization, quantitation and purification of deteriorated proteins

Methods for the study of protein deteriorations should have increasing attention. Usually, methods have been developed in response to a particular problem and not as part of a program in their own right. Obviously, a method for studying deteriorated proteins cannot be developed until something is known about the deterioration for which the method is to be developed. From the broad approach, consideration should be given to general methods for detecting all types of deterioration and, certainly, for the separation and partial purification of the products to allow for better study.

Acknowledgments

The author would like to thank Chris Howland for editorial assistance and Clara Robison for typing the manuscript. Background researches for this article were supported by FDA Grant FD00568 and NIH Grant AM26031.

Literature Cited

1. Anson, M. L. Adv. Protein Chem., 1945, 2, 361.
2. Whitaker, J. R. In "Food Proteins"; (Whitaker, J. R.; Tannenbaum, S. R., Eds.), Avi Publ. Co.: Westport, Conn., 1977, 14.
3. Lapanje, S. "Physicochemical Aspects of Protein Denatura-

tion"; John Wiley and Sons: New York, N.Y., 1978.
4. Van Holde, K. E. In "Food Proteins"; (Whitaker, J. R.; Tannenbaum, S. R., Eds.), Avi Publ. Co.: Westport, Conn., 1977, 1.
5. Chou, P. Y.; Fasman, G. D. Annu. Rev. Biochem., 1978, 47, 251.
6. Anfinsen, C. B.; Scheraga, H. A. Adv. Protein Chem., 1975, 29, 205.
7. Rose, G.; Winters, R. H.; Wetlaufer, D. B. FEBS Lettr., 1976, 63, 10.
8. Azari, P. R.; Feeney, R. E. J. Biol. Chem., 1958, 232, 293.
9. Azari, P. R.; Feeney, R. E. Arch. Biochem. Biophys., 1961, 92, 44.
10. Osuga, D. T.; Feeney, R. E. In "Food Proteins"; (Whitaker, J. R.; Tannenbaum, S. R., Eds.), Avi Publ. Co.: Westport, Conn., 1977, 209.
11. Greene, F. C.; Feeney, R. E. Biochemistry, 1968, 7, 1366.
12. Yeh, Y.; Iwai, S.; Feeney, R. E. Biochemistry, 1979, 18, 882.
13. Chen, Y.-H.; Yang, J. T. Biochem. Biophys. Res. Commun., 1971, 44, 1285.
14. Greenfield, N.; Fasman, G. D. Biochemistry, 1969, 8, 4108.
15. Uy, R.; Wold, F. Science, 1977, 198, 890.
16. Means, G. E.; Feeney, R. E. "Chemical Modification of Proteins"; Holden-Day: San Francisco, 1971.
17. Feeney, R. E.; Blankenhorn, G.; Dixon, H. B. F. Adv. Protein Chem., 1975, 29, 135.
18. Hodges, J., this publication.
19. Liener, I. E. In "Food Proteins. Improvement through Chemical and Enzymatic Modification"; (Feeney, R. E.; Whitaker, J. R., Eds.), American Chemical Society: Washington, D. C., Adv. Chem. Series, 1977, 160, 283.
20. Vaughan, D. A. In "Evaluation of Proteins for Humans"; (Bodwell, C. E., Ed.), Avi Publ. Co.: Westport, Conn., 1977, 255.
21. Feeney, R. E.; Clary, J. J.; Clark, J. R. Nature, 1964, 201, 192.
22. Feeney, R. E.; Abplanalp, H.; Clary, J. J.; Edwards, D. L.; Clark, J. R. J. Biol. Chem., 1963, 238, 1732.
23. Lightbody, H. D.; Fevold, H. L. Adv. Food Res., 1948, 1, 149.
24. Kline, L.; Gegg, J. E.; Sonoda, T. T. Food Technol., 1951, 5, 181.
25. Kline, L.; Hanson, H. L.; Sonoda, T. T.; Gegg, J. E.; Feeney, R. E.; Lineweaver, H. Food Technol., 1951, 5, 323.
26. Chio, K. S.; Tappel, A. L. Biochemistry, 1969, 8, 2827.
27. Saxena, V. P.; Wetlaufer, D. B. Biochemistry, 1970, 9, 5015.
28. MacDonnell, L. R.; Lineweaver, H.; Feeney, R. E. Poultry Sci., 1951, 30, 856.

29. Sjoberg, L. B.; Feeney, R. E. Biochim. Biophys. Acta, 1968, 168, 79.
30. Schöberl, P.; Rambacher, P. Annalen Chemie, 1939, 538, 84.
31. Feeney, R. E.; MacDonnell, L. R.; Ducay, E. D. Arch. Biochem. Biophys., 1956, 61, 72.
32. Nashef, A. S.; Osuga, D. T.; Lee, H. S.; Ahmed, A. I.; Whitaker, J. R.; Feeney, R. E. J. Agric. Food Chem., 1977, 25, 245.
33. Whitaker, J. R., this publication.
34. Packer, L.; Kellogg, E. E., III, this publication.
35. Rogers, T. B.; Gold, R. A.; Feeney, R. E. Biochemistry, 1977, 16, 2299.
36. Rogers, T. B. "The Chemistry of Iron and Anion Binding by Transferrins"; Ph.D. Thesis, University of California, Davis, 1977.
37. Schroeder, D. D.; Allison, A. J.; Buchanan, J. M. J. Biol. Chem., 1969, 244, 5856.
38. Feeney, R. E. In "Evaluation of Proteins for Humans"; (Bodwell, C. E., Ed.), Avi Publ. Co.: Westport, Conn., 1977, 233.
39. McKinney, L. L.; Weakley, F. B.; Eldridge, A. C.; Campbell, R. E.; Cowan, J. C.; Picken, J. C., Jr.; Biester, H. E. J. Am. Chem. Soc., 1957, 79, 3932.
40. Feeney, R. E. In "Food Proteins. Improvement through Chemical and Enzymatic Modification"; (Feeney, R. E.; Whitaker, J. R., Eds.), American Chemical Society: Washington, D.C., Adv. Chem. Series, 1977, 160, 3.
41. Rando, R. R. Science, 1974, 185, 320.
42. Rando, R. R. Acc. Chem. Res., 1975, 8, 281.
43. Abeles, R. H.; Maycock, A. L. Acc. Chem. Res., 1976, 9, 313.
44. Metcalf, B., this publication.
45. Feinstein, G.; Feeney, R. E. J. Biol. Chem., 1966, 241, 5183.
46. Ryan, D. S.; Feeney, R. E. In "Bayer Symposium V, Proteinase Inhibitors"; (Fritz, H.; Tschesche, H.; Greene, L. J.; Truscheit, E., Eds.), Springer-Verlag: Berlin, 1974, 378.
47. Means, G. E.; Ryan, D. S.; Feeney, R. E. Acc. Chem. Res., 1974, 7, 315.
48. Eshdat, Y.; McKelvy, J. F.; Sharon, N. J. Biol. Chem., 1973, 248, 5892.
49. Eshdat, Y.; Dunn, A.; Sharon, N. Proc. Nat. Acad. Sci. USA, 1974, 71, 1658.
50. Walsh, R. G.; Nashef, A. S.; Feeney, R. E. Intern. J. Peptide Protein Res., 1979, in press.
51. Feeney, R. E.; Osuga, D. T. In "Methods of Protein Separation 1"; (Catsimpoolas, N., Ed.), Plenum Press: New York, N.Y., 1975, 127.
52. Catsimpoolas, N., Ed., "Methods of Protein Separation 1"; Plenum Press: New York, N.Y., 1975.

53. Catsimpoolas, N., Ed., "Methods of Protein Separation 2"; Plenum Press, New York, N.Y., 1976.
54. Creighton, T. E. "Abstracts of Papers," XIth Intern. Congress Biochem. Toronto, Canada, 1979, 140.
55. Masters, P. M.; Friedman, M., this publication.
56. Nashef, A. S.; Osuga, D. T.; Feeney, R. E. Anal. Biochem., 1977, 79, 394.
57. Rucker, R. B.; Tinker, D. Intern. Rev. Exper. Pathol., 1977, 17, 1.
58. Rucker, R. B., this publication.
59. Moldave, K.; Harris, J.; Sabo, W.; Sadnik, I. Fed. Proc., 1979, 38, 1979.
60. Bada, J. L.; Kvenvolden, K. A.; Peterson, E. Nature, 1973, 245, 308.
61. Masters, P. M.; Bada, J. L.; Zigler, J. S., Jr. Nature, 1977, 268, 71.
62. Bada, J. L.; Schroeder, R. A. Naturwissenschaften, 1975, 62, 71.
63. Fennema, O., Ed., "Behavior of Proteins at Low Temperatures"; American Chemical Society: Washington, D.C., Adv. Chem. Series, 1979, in press.
64. Matsumoto, J. J., this publication.
65. Fukushima, D., this publication.
66. Feeney, R. E.; Hill, R. M. Adv. Food Res., 1960, 10, 23.

RECEIVED October 18, 1979.

Posttranslational Chemical Modification of Proteins

ROSA UY
3M Central Research Lab, St. Paul, MN 55101

FINN WOLD
Department of Biochemistry, University of Minnesota, St. Paul, MN 55108

It is now well established that the poly-(amino acid) chains assembled from the 20 amino acids specified by the genetic code undergo extensive processing before the biologically active final products of protein biosynthesis are obtained. Most of this processing involves covalent modification, either by making or breaking peptide bonds or by derivatizing the free α-NH_2 or α-COOH groups or the amino acid side chains. These covalent modifications, which may take place immediately after the formation of the amino acyl-tRNA, during the elongation of the nascent polypeptide chain on the polysomes or after the polymer has been completed and is being transported through intracellular or extracellular space, can all be considered as posttranslational modifications. A recent refinement on this terminology has been to distinguish between cotranslational modifications (all events taking place during polymerization on the polysomes) and post-translational modification (all events after the release of the protein precursor from the polysomes). In this system of designation, the amino acid modifications that take place at the level of amino acyl-tRNA (e.g. the formation of N-formyl Met-tRNA) should presumably be referred to as pretranslational modifications. There is probably some merit in this system from the point of view of economy of communication, but it is also probably safe to say that at the current state of knowledge the assignment of a given modification reaction to pre-, co- or posttranslational status will be mostly speculative, and in this paper the single designation posttranslational will be used for all these reactions.

The biological function of many of these posttranslational modification reactions is also tenuous at this stage, even if we can rationalize some of them as components of well understood processes. The most obvious example of such established processes is the oxidation of sulfhydryl groups resulting in the formation of disulfide bridges which are essential structural

0-8412-0543-4/80/47-123-049$05.00/0

features of most extracellular native proteins. Similarly, in
the case of rigid proteins such as collagen and elastin fibers,
several chemical reactions consisting of hydroxylation, oxidation, and deamination occur. These subsequently lead to the
characteristic covalent crosslinking which provides the molecular
basis for the structural and elastic properties of connective
tissues (1). It has also been established that peroxidase-
catalyzed halogenation of thyroglobulin is an essential step in
the biosynthesis of thyroxine, and that several of the reversible
modification steps (e.g. phosphorylation) are involved in the
regulation of metabolic reactions through the activation or inactivation of enzymes and regulatory proteins (histones). Finally,
recent studies in the area of glycoprotein structure and function
suggest an important role of the glycosylation reactions in
biological communications.

   Even when the purpose of a given posttranslational modification is understood, an examination of how and where it occurs is
also likely to yield only limited information. The cell biological sites and processes involved in the reactions, the specificity by which certain amino acid residues or specific peptide
bonds are selected for chemical modification and the mechanism by
which the transformations are carried out remain obscure for a
large number of these reactions.

   The field of posttranslational modification of proteins has
been reviewed (2, 3), and the main purpose of this article is to
bring earlier reviews up to date, and to explore some of the most
recent developments in the field. Figure 1 summarizes our
current knowledge of posttranslation modification reactions.
Since the detailed listing of most of these reactions has been
reported before, the reader is referred to the original tabula-
tion for proper nomenclature and literature references.

   If one considers all known covalent alterations of the polyamino acid chains produced by living cells, the total number and
types of reaction become rather unwieldy, and it is useful to
classify them into three broad groups: 1) modification (cleavage
and formation) of the peptide bond, 2) modification of the
terminal $\alpha-NH_2$ and $\alpha-COOH$ groups and 3) modification of amino
acid side chains. The first of these three will not be considered in this discussion beyond the reminder that limited
proteolysis is an essential and very broadly observed phenomenon
in all living systems. Ever since it was realized that, although
the initiation of protein synthesis requires N-formylmethionine
as the N-terminal residue the finished product is rarely, if ever,
found with that residue still attached, it was clear that most,
perhaps even all, biologically active proteins have undergone
proteolytic cleavage. The recent signal peptide hypothesis has
added another dimension to this processing step in proposing that
a hydrophobic N-terminal sequence demonstrated for a large number
of protein precursors is essential in directing a given protein
to or through the membranes of the cell. This signal peptide is

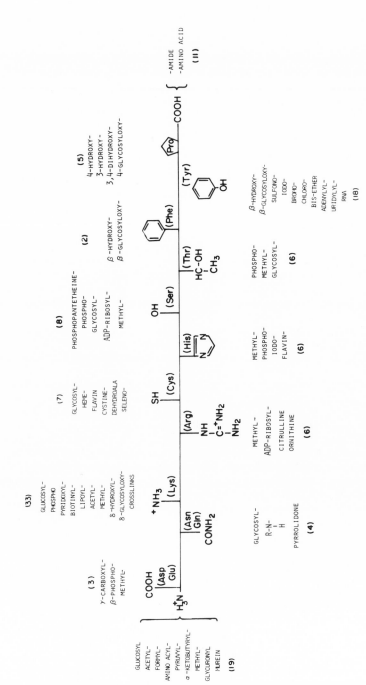

Figure 1. These known derivatives are summarized from Ref. 2, the figures in the parentheses refer to number of the known identified derivatives. Four new derivatives have since been added; saleno-Cys (55), RNA-Tyr (56), and for glycosylation of Val and Lys (see Ref. 53 and 54.)

subsequently removed in the processing of the precursor to yield
the final protein product. In addition to these virtually uni-
versal peptide bond cleavage reactions, a number of specific
biological processes also involve limited proteolytic cleavage.
Notable examples of such processes are activation of zymogens
and prohormones (4, 5), virus and ribosome assembly (6), the
blood coagulation cascade (7), and complement activation (8).

Modification of N- and C-Terminal Amino Acid Residues

The second category of reactions, those involving modifi-
cation of the terminal α-NH$_2$ and α-COOH groups of proteins, contains a relatively short list of different derivatives. The
carboxy-terminal end of some proteins (mostly relatively low
molecular weight physiologically active hormones and venoms) has
been found to be an amide rather than a free α-carboxylate
function; in other proteins the α-COOH is found to be methylated
and recently it has been reported that the amino acid tyrosine
can be added to the C-terminal end of tubulin (9). The latter
reaction requires ATP and the new derivatized C-terminus, containing a regular α-amino acid, just like any protein, can only
be recognized as a posttranslational modification, if the precursor had first been characterized or if the DNA sequence had
been established not to include the codon for the C-terminal
tyrosine.

The N-terminal α-NH$_2$ group appears to be involved in more
modification than does the C-terminus. Acetylation, formylation
and methylation are three quite common N-terminal derivatives,
and just like in the case of the C-terminal end, amino acids can
be added to the N-terminus of the finished polypeptide chain in
the absence of ribosomes, but in this case the donor is amino
acyl-tRNA.

One interesting modification at the N-terminal end is the
reaction by which the protein acquires an α-keto acyl terminus
(10-13). In the processing of pro-histidine decarboxylase as
shown in Figure 2, an interior Ser-Ser bond can be cleaved,
leading to the formation of two polypeptide chains, a β subunit
with serine as its C-terminal and an α subunit with serine as its
N-terminal end. This latter serine is presumably first converted
to dehydroalanine through dehydration, and subsequently, the
release of one mole of ammonia results in the formation of a
pyruvoyl group on the N-terminal of the α subunit. There is also
a possibility that the dehydration of serine (or dehydrosulfation
of cystine) can take place internally without peptide bond
cleavage). Dehydroalanine itself has actually been observed as a
prosthetic group in enzymes (14). The N-terminal of urocanase
is presumably modified in a similar fashion from threonine to
form α-ketobutyrate through a dehydroaminobutyrate intermediate
(15, 16). The function of pyruvate and α-ketobutyrate appears
to provide an essential carbonyl group for activity. Similar

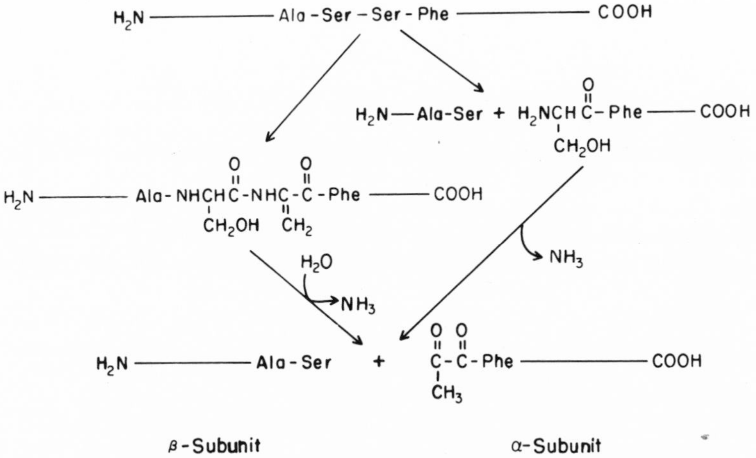

Figure 2. Activation of pro-histidine decarboxylase

carbonyl functions can be provided by pyridoxal phosphate in
other enzymes.

The direct acylation and methylation of the free $\alpha-NH_2$
groups of proteins have been proposed to be useful in providing
resistance toward proteolytic attacks. Although the basis for
this explanation is not always readily apparent from the known
specificities of proteases, it may be valid in some cases. Thus,
the acetylated N-terminus of α-crystallin, the major protein
found in eye lens (17), is presumably important for the protein
to survive in an environment rich in leucine aminopeptidase. On
the other hand, it is difficult to rationalize that the acetylated N-terminus of bovine pancreatic α-amylase is in any way
responsible for the fact that the enzyme is exceedingly stable
against tryptic and chymotryptic digestion (18). The function of
the acetylation is, in this case, as obscure as is the basis on
which α-amylase is selected for acetylation among the many non-
acetylated companion pancreatic proteins.

The mechanism of N-acetylation of α-crystallin is quite
interesting. The N-terminal residue has been identified as N-
acetyl methionine. This methionine residue is derived from Met-
$tRNA_f^{met}$ which is responsible for the initiation of the polypeptide chain and not Met-tRNA^{met} which incorporates the methionine
residue in the growing polypeptide chain. It is clear that the
N-acetylation is a true posttranslational process since acetyl
Met-tRNA cannot replace formyl Met-tRNA$_f^{met}$. Moreover, N-acetylation occurs when the polypeptide chain has reached a length consisting of approximately 25 amino acid residues. Other proteins,
such as ovalbumin, are also acetylated during the early stages of
polymerization on the polysome, and the protein acetyltransferase
activity must therefore be associated with the protein-synthe-
sizing apparatus.

Modification of Amino Acid Side Chains

The third category of posttranslational reactions, those
involved in covalent modification of the amino acid side chains,
is by far the largest. According to the data in Fig. 1 (and Ref.
2) there are about 98 known derivatives of amino acid side chains
in proteins. In the following paragraphs some of these will be
discussed briefly.

A number of proteins such as histones, cytochrome c and
certain flagellar proteins are found to contain methylated amino
acids (19). Three different methylases have been characterized
and all require S-adenosylmethionine (SAM) as the methyl donor
(20-23). Protein methylase I (SAM-protein arginine methyl transferase) methylates the guanidine side chain of arginine residue;
protein methylase II (SAM-protein carboxyl methyltransferase)
transfers methyl groups only to β- and γ-carboxyl groups in the
peptide chain. Carboxyl groups in the α position cannot serve as
acceptors. Protein methylase III (SAM-protein lysine

methyltransferase) is involved in the methylation of the N^ϵ of
the lysine residue. The presence of monomethyl, dimethyl and
trimethyl lysine suggests that methylation of lysine may involve
more than one enzyme. An S-adenosylmethionine:protein-lysine
transferase was purified recently from Neurospora crassa (24).
The enyzme recognized the sequence X-lys-lys-Y, where X and Y can
vary but the two adjacent lysines in the sequence are an absolute
requirement. The length of the peptide acting as substrate is
also important.

A methylesterase which catalyzes the hydrolysis of γ-glutamyl methyl esters of membrane bound proteins in Salmonella
typhimurium and E. coli has recently been identified (25).
Apparently, these membrane-bound proteins undergo methylation by
a S-adenosylmethionine requiring methyltransferase (similar to
transferase II). In this case the methylation and demethylation
are directly associated with the chemotactic mobility of the
microorganisms. When the cells are exposed to a chemotactic
attractant, methylation of the membrane-bound proteins increases
and straight-line movement up the gradient is induced. When the
attractant is removed or a repellent substituted, the esterase
decreases the methylation and random movement results. These
control mechanisms are analogous in regulation to some of the
reversible processes such as adenylation (26), uridylation (27)
and phosphorylation (28).

The molecular basis for regulation of enzymatic activity
through phosphorylation and dephosphorylation has been established in many enzyme systems (29). The significance of these
reactions in histones, ribosomal proteins and RNA polymerase is
not known. In an attempt to establish the specificity of the
cyclic AMP-dependent protein kinases, the structure of several
substrates have been determined (30). The data indicate that the
sequence around the phosphorylated serine residue all contain two
basic amino acids separated by no more than two residues from the
N-terminal of the susceptible serine (e.g. -Arg-Arg-X-Y-Ser-).
In synthetic peptides, the presence of the two adjacent basic
amino acids is an essential requirement for the peptide to be
phosphorylated with kinetic constants comparable to those of the
physiological substrates (31). It has also been observed that in
addition to the requirement for the two adjacent basic residues,
the nature of the amino acid C-terminal to the phosphoserine is
involved as a specificity determinant. When serine was replaced
by threonine, the synthetic peptides were no longer active as
substrate.

A fairly recent and fascinating family of reactions is the
incorporation of ADP-ribosyl into proteins as shown in Figure 3.
In the ADP-ribosylation of E. coli DNA dependent RNA polymerase
upon infection with bacteriophage T4 (32, 33), the reaction
requires NAD^+ and proceeds with the concomitant release of nico-
tinamide and a proton. The effect of the modification reaction
is to turn off host transcription. Two arginine residues were

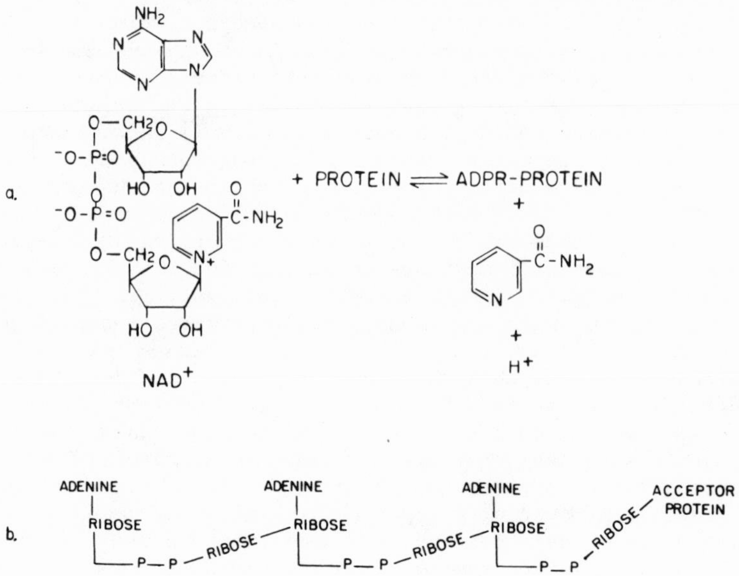

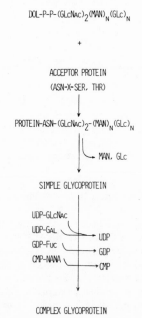

Figure 3. a. ADP-ribosylation of protein; b. Formation of poly-(ADP-ribose)-protein

DOL-P-P-(GLcNAc)$_2$(MAN)$_N$(GLc)$_N$

+

ACCEPTOR PROTEIN
(ASN-X-SER, THR)

PROTEIN-ASN-(GLcNAc)$_2$-(MAN)$_N$(GLc)$_N$

MAN, GLc

SIMPLE GLYCOPROTEIN

UDP-GLcNAc
UDP-GAL   UDP
GDP-FUC   GDP
CMP-NANA   CMP

Figure 4.   General scheme for the bio-synthesis of Asn-linked glycoprotein

COMPLEX GLYCOPROTEIN

found to be involved in the link to the ADP-ribose through a C–N bond (34, 35). In contrast to phosphorylation, where a specific primary amino acid sequence is recognized during modification, ADP-ribosylation appears to have broad specificity. The ADP-ribose linkage found in elongation factor a after exposure to diptheria toxin has been proposed to involve a linkage to a basic residue such as histidine, lysine or tyrosine (36–39). Repeating units of ADP-ribose linked together by ribose residues which form poly-(ADP-ribose) have been established to be the modifying reagent in the case of histones. Based on the instability of the poly-(ADP-ribose)-protein towards neutral hydroxylamine treatment and alkali, an ester linkage involving the carboxyl group of either glutamate or an aspartate residue has been proposed to be the site of modification (40). In histones F1(H1), the ADP-ribosyl group was found to be linked through a serine residue (41).

The biological function of ADP-ribosylated nuclear proteins is not clear. Poly-(ADP-ribose) synthetase has been reported to be stimulated by histones, but histones themselves do not serve as acceptors in vitro (42). The endogenous acceptor in vivo might involve proteins other than histones. Recently an enzyme that cleaves the ADP-ribosyl histone linkage has been purified from rat liver (43).

Protein glycosylation is receiving a good deal of attention at present both as a well-defined example of posttranslational modification, and as an important step in key biological processes leading to specific links of cell-cell and cell-molecule communication. The mechanism of synthesis of the asparagine-linked oligosaccharides has been studied extensively. Apparently, glycosylation occurs in a multistep process as shown in Figure 4. The oligosaccharide moiety is first built up on a lipid carrier (-pyrophosphoryl-dolichol) and is then transferred to the asparagine residue of the newly synthesized peptide. The glycopeptide is further modified by the removal and addition of monosaccharides. It has been demonstrated that glucose is present in the lipid-oligosaccharide and is subsequently trimmed off during the processing (44). Final attachment of terminal sugars such as sialic acid and fucose residues presumably occurs in the Golgi apparatus where the specific enzymes are localized in high concentration.

An examination of the specificity of the Asn-linked glycoproteins indicate that the sequence Asn-X-Ser (Thr) is an essential feature for glycosylation. Various proteins such as ovalbumin, α-lactalbumin and RNAse A which have this common sequence serve as substrates but in some instances only after denaturation (45). RNAse A can be converted to a species which is electrophoretically identical to RNAse B (the natural glycosylated form of the enzyme). The length of the Asn-X-Ser sequence does not seem to affect the rate of glycosylation. The (acetylated) tripeptide itself is as good a substrate as longer peptides which contain the specific tripeptide residue, provided the asparaginyl residue does not have a free α-amino group (46). The fact that denaturation may be required before a protein becomes an acceptor for the incorporation of the oligosaccharide to the Asn-X-Ser (Thr) region suggests that three-dimensional folding determines the accessibility of this sequence to the specific oligosaccharide transferase, and it is therefore quite likely that the initial glycosylation occurs after polymerization and folding of the protein into its final conformation.

The specificity and mechanism of O-glycosylation have not been clearly established. A comparison of the sequences around the 28 known O-glycosidically substituted seryl and threonyl residues in ovine submaxilliary mucin did not show any homologies (47).

It was originally thought that the glycosyl moiety of the glycoproteins provided the marker which distinguishes the proteins synthesized for secretion from those synthesized for intracellular use (48). This idea is no longer valid. In bovine pancreatic juice, for example, only ribonuclease B, C, D and deoxyribonuclease are secreted as glycoproteins. From the nutritional point of view, approximately 60% of the secreted bovine milk proteins are glycosylated and in human plasma, although the majority of the proteins are glycosylated, these represent only 50% of the total protein mass. There is also increasing evidence suggesting that the glycosyl moiety can serve as a "built-in" signal for intra- and intercellular communications.

Modification by Non-Enzymatic Reactions

Proteins can also undergo spontaneous chemical modification through changes in the external environment. The oxidation and reduction of SH groups can occur readily and some of the cross-linking reactions in connective tissues can also occur spontaneously without the participation of an enzyme. Under alkaline conditions, the formation of dehydroalanine is favored through the β-elimination of serine, cystine or cysteine residues. The hydrolysis of γ-glutamyl methyl ester can probably occur without the presence of methylesterase since this derivative is quite labile at neutral pH, and glutamine and asparagine will also slowly hydrolyze under physiological conditions.

The rates of nonenzymatic deamination of glutaminyl and asparaginyl residues have been examined in 42 peptides (49). It was demonstrated that the intramolecular steric hindrance and the charge on residues in close proximity to the glutaminyl or asparaginyl residue play an important role in determining the rate of the deamination reaction. During the aging of α-crystallin, asparagine is progressively deaminated to aspartic acid. This modification step does not seem to be mediated by a specific enzyme since deamination can be induced in vitro by incubating pure α-crystallin in solution. The deamination reaction in α-crystallin might be a phenomenon of aging.

Although several enzymes which catalyze the conversion of glutamic acid to pyrrolidone carboxylic acid have been identified (50), pyrrolidone carboxylic acid is also formed nonenzymatically from free glutamine or from glutamine-terminated peptides and proteins during prolonged incubation. In aqueous solutions, the formation of pyrrolidone carboxylic acid depends on the concentration of the ionic species of glutamic acid. At 37°, 10% of the glutamine is converted to pyrrolidone carboxylate in 24 hours. Peptides that possess glutamine in the N-terminal position exhibit a lability similar to that of glutamine (51).

The ADP-ribosylation of proteins is also feasible in the absence of the specific enzymes. Covalent adducts of polylysine, bovine serum albumin, lysine rich histone (F1) and DNase with ADP-ribose and ribose-5-phosphate have been prepared at pH 7.4 and 9.5 (52). The formation of a Schiff base was indicated by the incorporation of ^{3}H into the adducts upon treatment with NaB^{3}H$_4$.

In a similar reaction, incubating proteins in the presence of glucose or lactose leads to the direct attachment of glucose to amino groups. Hemoglobin A$_{IC}$, a naturally-occurring hemoglobin derivative, is presumably formed this way from hemoglobin A through the glycosylation of the N-terminal valine of the β chain (53). The reaction is irreversible and the rate is a function of the environment (glucose concentration) of the erythrocytes. The addition of the sugar group to the hemoglobin molecule occurs non-enzymatically through the reaction of the aldehyde group of the glucose to form a Schiff base adduct with the N-terminus. The linkage subsequently undergoes an Amadori rearrangement as shown in Figure 5.

Several different glycolytic intermediates such as glucose-6-phosphate, fructose-6-phosphate, fructose 1,6-diphosphate and glyceraldehyde-3-phosphate were proposed to serve as substrates. The direct reaction of hemoglobin with glucose is much slower and less specific than that with glucose-6-phosphate.

Recently, a similar glycosylation has also been reported in human serum albumin. Approximately 6-15% of the albumin was found to exist naturally in the glycosylated form (54). In this case lysine residues rather than just the N-terminal α-NH$_2$ groups were shown to be involved in the reaction. Since the extent of these reactions appears to reflect the glucose concentration in the environment of the glycosylated proteins, their use as diagnostic indicators in diabetes is being explored.

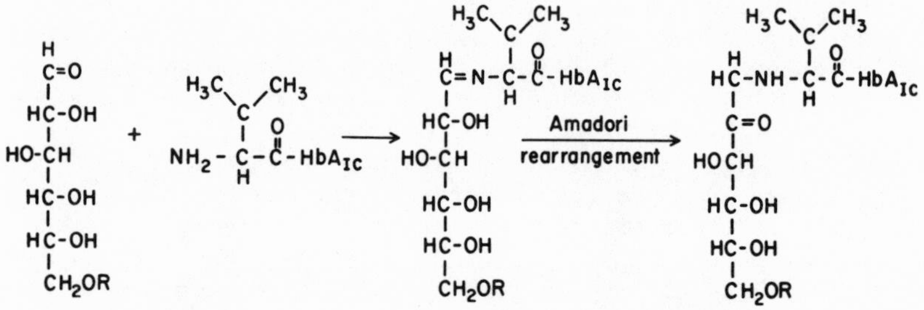

Figure 5.   The reaction of glucose with hemoglobin

Literature Cited

1. Gallop, R. M.; Blumenfeld, O. D.; Seifter, S., Annu. Rev. Biochem., 1972, 41, 617.
2. Uy, R.; Wold, F., Science, 1977, 198, 890.
3. Whitaker, J. R., in "Food Proteins, Improvement through Chemical and Enzymatic Modifications", Feeney, R. E.; Whitaker, J. R., eds., Amer. Chem. Soc., Washington, D.C., 1977.
4. Neurath, H.; Walsh, K. A., Proc. Natl. Acad. Sci. U.S.A., 1976, 73, 4825.
5. Tager, H. S.; Steiner, D. F., Annu. Rev. Biochem., 1974, 43, 567.
6. Hershko, A.; Fry, M., Annu. Rev. Biochem., 1975, 44, 775.
7. Davie, E. W.; Kerby, E. P., Curr. Top. Cell. Regul., 1973, 7, 51.
8. Muller-Eberhard, H. J., Harvey Lect., 1971, 66, 75.
9. Raybin, D.; Flavin, M., Biochemistry, 1977, 16, 2189.
10. Snell, E. E., Trends in Biochem. Sci., 1977, 2, 131.
11. Demetriou, A. A.; Cohn, M. S.; Tabor, C. W.; Tabor, H., J. Biol. Chem., 1978, 253, 1684.
12. Seto, B., J. Biol. Chem., 1978, 253, 4525.
13. Satre, M.; Kennety, E. P., J. Biol. Chem., 1978, 253, 479.
14. Havir, E. A.; Hansen, K. R., Biochemistry, 1975, 14, 1620.
15. Lynch, M. C.; Phillips, A. T., J. Biol. Chem., 1972, 247, 7799.
16. Kapke, G.; Davis, L., Biochemistry, 1975, 14, 4273.
17. Bloemendal, H., Science, 1977, 197, 127.
18. Uy, R.; Wold, F., Fed. Proc., 1977, 36, 700.
19. Cantoni, G. L., Annu. Rev. Biochem., 1975, 44, 435.
20. Paik, W. K.; Kim, S., J. Biol. Chem., 1970, 245, 6010.
21. Paik, W. K.; Kim, S., J. Biol. Chem., 1968, 243, 2108.
22. Kim, S., Arch. Biochem. Biophys., 1973, 161, 652.
23. Kim, S., Arch. Biochem. Biophys., 1973, 157, 476.
24. Durben, E.; Nochumson, S.; Kim, S.; Paik, W. K., J. Biol. Chem., 1968, 253, 1427.
25. Stock, J. B.; Koshland, D. E., Proc. Natl. Acad. Sci. U.S.A., 1978, 75, 3659.
26. Heinrikson, R. L.; Kingdon, H. S., J. Biol. Chem., 1971, 246, 1099.
27. Adler, S. P.; Purich, D.; Stadtman, E. R., J. Biol. Chem., 1975, 250, 6264.
28. Rubin, C. S.; Rosen, O. M., Annu. Rev. Biochem., 1975, 44, 831.
29. Segal, H. L., Science, 1973, 180, 25.
30. Yeaman, S. J.; Cohen, P.; Watson, D. C.; Dixon, G. H., Biochem. J., 1977, 162, 411.

31.   Zetterqvist, O.; Ragnarsson, U.; Humble, E.; Berglund, L.; Engstrom, L., Biochem. Biophys. Res. Commun., 1976, 70, 696.

32.   Hayaishi, O.; Ueda, K., Annu. Rev. Biochem., 1977, 46, 95.

33.   Hayaishi, O., Trends in Biochem. Sci., 1976, 1, 9.

34.   Goff, C. G., J. Biol. Chem., 1974, 249, 6181.

35.   Rohres, H.; Zillig, W.; Mailhammer, R., Eur. J. Biochem., 1975, 60, 227.

36.   Honjo, T.; Hayaishi, O., Curr. Top. Cell. Regul., 1973, 7, 87.

37.   Honjo, T.; Ueda, K.; Tanabe, T.; Hayaishi, O., in Second International Symp. Metab. Interconvers. Enzymes. Wieland, O., Helmreich, E., Holzer, H., eds., Berlin Springer, 1972, p. 193.

38.   Robinson, E. A.; Henriksen, O.; Maxwell, E. S., J. Biol. Chem., 1974, 249, 5088.

39.   Edson, C.; Ueda, K.; Hayaishi, O., J. Biochem., (Tokyo), 1975, 77, 9.

40.   Hayaishi, O., in The 11th Miami Winter Symposium, 1979, Jan.

41.   Ord, M. G.; Stocken, L. A., Biochem. J., 1968, 161, 583.

42.   Edson, C. M.; Okayama, H.; Fukushima, M.; Hayaishi, O., Fed. Proc., 1976, 35, 1722.

43.   Okayama, H.; Honda, M.; Hayaishi, O., Proc. Natl. Acad. Sci. U.S.A., 1978, 75, 2254.

44.   Tabas, I.; Schesinger, S.; Kornfeld, S., J. Biol. Chem., 1978, 253, 716.

45.   Pless, D. D.; Lennarz, W. J., Proc. Natl. Acad. Sci. U.S.A., 1977, 74, 134.

46.   Struck, D. K.; Lennarz, W. J.; Brew, K., J. Biol. Chem., 1978, 253, 5786.

47.   Hill, H. D., Jr.; Schwyzer, M.; Steinman, H. M.; Hill, R. L., J. Biol. Chem., 1977, 252, 3799.

48.   Winterburn, P. S.; Phelps, C. F., Nature, 1972, 236, 147.

49.   Robinson, A. B.; Scotchler, J. W.; McKerrow, J. H., J. Amer. Chem. Soc., 1973, 95, 8156.

50.   Orlowski, M.; Meister, A., "The Enzymes", Boyer, P. D., ed., Academic Press, N.Y., 1971, 123.

51.   Blomback, B., Methods in Enzymology, 1967, 11, 398.

52.   Kun, E.; Chang, A. C. Y.; Sharman, M. L.; Ferro, A. M.; Nitecki, D., Proc. Natl. Acad. Sci. U.S.A., 1976, 73, 3131.

53.   Cerami, A.; Koenig, R., Trends in Biochem. Sci., 1978, 3, 73.

54.   Day, J. F.; Thorpe, S. R.; Baynes, J. W., J. Biol. Chem., 1979, 254, 595.

55.   Tanaka, H.; Stadtman, T.C., J. Biol. Chem., 1979, 254, 447-452.

56.   Ambrose, V.; Baltimore, D., J. Biol. Chem., 1978, 253, 5263-5266.

RECEIVED October 26, 1979.

Chemical Changes in Elastin as a Function of Maturation

ROBERT B. RUCKER and MICHAEL LEFEVRE

Department of Nutrition, University of California, Davis, CA 95616

This review focuses upon the post-translational modification
and chemical changes that occur in elastin. Outlined are the
steps currently recognized as important in the assembly of profibrillar elastin subunits into mature fibers. Descriptions of
some of the proposed mechanisms that appear important to the
process are also presented. It will be emphasized that from the
standpoint of protein deterioration, elastin is a very novel
protein. Under normal circumstances, the final product of elastin
metabolism, the elastin fiber does not undergo degradation that
is easily measured. Unlike the metabolism of many other proteins,
deterioration or degradation is most evident biochemically in the
initial stages of synthesis rather than as a consequence of
maturation. Since the presence of crosslinks is an essential
component of mature elastin, a section of this review also
addresses important features of crosslink formation.

For purposes of definition, we will use the following terms
to designate the various forms of elastin. The term, non-crosslinked elastin, will be used as a general description for proposed precursors to mature elastin that appear to be rapidly
modified during the initial stages of elastic fiber formation.
With respect to one of these precursors, the term, tropoelastin,
has been used as a designation for a non-crosslinked elastin
precursor of approximately 70,000 daltons (1). Since it is
currently the best characterized of the non-crosslinked elastins
and is used extensively by those familiar with elastin, this
term will be retained. Elastin will be used to designate the
protein in its crosslinked form. This term, however, is at best
operational, since elastin is only isolated from tissues or cell
culture by procedures that would be offensive to most protein
chemists. As a component of extracellular matrices, elastin is
extremely insoluble and in close association with many other
extracellular components (2). In order to remove these components,
harsh treatments such as autoclaving, extraction with alkali or

0-8412-0543-4/80/47-123-063$05.00/0
© 1980 American Chemical Society

organic acids, extraction with non-elastolytic enzymes, or repeat-
ed extraction with denaturants are required (1,3-7). Finally,
when the term, elastic fiber is used, the context will also be
operational, since elastic fibers appear to be composed of
several components (elastin, collagen and other fibrillar
proteins) in varying amounts (2).

Models for Elastin and Elasticity

Before describing the major steps in elastin biosynthesis, a
general consideration of the function of elastin is pertinent.
When hydrated, elastin fibers possess some of the mechanical properties of polymeric rubbers (1,2,8,9). Skin, lung, ligaments,
and major blood vessels contain a high concentration of elastin,
because of the need in these tissues for long-range, reversible
extensibility (9). When elastic fibers are observed, they are
often found branched and fused in the form of a complex network.
However, there is a degree of order in tissues that are subjected
to unidirectional stress, such as the ligamentum nuchae of un-
gulates, with the orientation of the fibers parallel to the direction of stress. In major blood vessels, elastic fibers take
on a lamellar arrangement in the form of concentric sheets (10,11).
Serafini-Fracassini et al. (12) and Cleary and Cliff (13)
have recently proposed that elastin in fibers appears to be
present as filaments (15-25Å in diameter). Serafini-Fracassini
et al. (12) argue that only the polymeric chains making up the
filaments are crosslinked. There is now sufficient data to
suggest that the filaments may contain a degree of ordered structure (1,8,12-14). The ordered structure, however, is quite different from that for other structural proteins, such as collagen.
The polymeric chains making up elastin appear to exist in the
form of β-spirals (β-turn structures) separated every 6000-8000
daltons by α-helical segments containing a high concentration of
crosslinks (14).
This is an important point, since a highly ordered network
for elastin would be incompatible with the entrophic interpreta-
tion of an elastic recoil; particularly the classical view which
requires that the crosslinked polymeric chains be in random conformation. Serafini-Fracasini et al. (12) argue that the filaments possess such small diameters that they may act, however, as
random chains. Morphological evidence suggests that the filaments
are bonded by non-covalent interactions to form a reticulum or
three-dimensional network as shown in Figure 1. We feel that the
filament model is attractive from a biological point of view,
since a degree of molecular organization would be expected in
order to form elastic fibers that interact intimately with the
other filamentous elements that comprise the extracellular matrix.
It should be noted, however, that other models have been proposed
for elastin. A description of these models may be found in reviews by Gosline (9) and Sandberg (1).

Special features of elastin structure are its unique amino acid composition (Table I) and amino acid sequences (Table II). Elastin is one of the most apolar proteins in nature. The high concentration of val-pro sequences as well as the crosslinks represented by amino acids such as desmosine (Des), isodesmosine (Ide) and lysinonorleucine (LNL) confer chemical stability ([1]). Considerable effort over the last few years has been directed toward sequencing portions of elastin. Data from these studies have provided much of the basis for the fibrillar model given in Figure 1. It is now well established that the polymeric chains of elastin are composed of alternating segments different in amino acid composition ([19-24]). What are viewed as the extendable segments contain repeating short peptide units characterized by their high content of valine, proline and glycine. Further, from data for amino acid sequences around some of the crosslinking amino acids ([19,20,23]) it is known that the peptides in these regions fall into two major categories based upon the amino acid residues following lysine. In one group of peptides, an aromatic amino acid residue is usually found adjacent to lysine, whereas in the other group of peptides an alanine is usually found. This is an important finding, because, as will be pointed out later, lysine is the precursor of the crosslinking amino acids in elastins. The presence or absence of an aromatic amino acid residue adjacent to lysine appears to determine whether or not it will be enzymatically modified. In addition, interactions involving aromatic rings may facilitate the transfer of electrons in the ultimate oxidation or reduction of certain crosslinking amino acids ([19-21],23).

Biosynthesis of Elastin

Elastic fibers are usually found in tissues rich in smooth muscle or tissues containing fibroblasts possessing some of characteristics of smooth muscle cells ([4]). There is a recent report, however, that suggests elastin-like proteins may be secreted from chondrocytes ([25]). When elastin is secreted, it is accompanied by other proteins that appear to be important to its alignment into fibrils. One of these proteins is referred to as microfibrillar protein (cf. Table I, ref. [2]). When elastin is secreted, it combines with the microfibrillar protein to form a complex which is initially rich in the microfibrillar protein. The ratio of microfibrillar protein to elastin, however, appears to decrease upon maturation ([2]). Other proteins are also secreted with microfibrillar protein and elastin. It is now clear that bound to elastin in its non-crosslink form(s) is a trypsin-like neutral proteinase ([26]). This proteinase effects

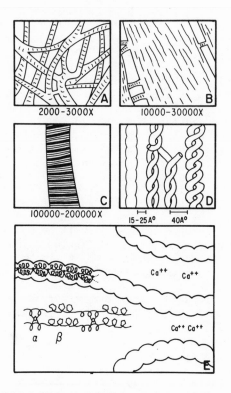

Figure 1. A model for elastin. The basis for the various figures are taken from references 8 and 12–18. At 2000–3000 magnification, mature elastin fibers appear rope-like as shown in A. The fibers are often branched and interconnected. At higher magnification, the fibers appear amorphous and smaller interconnecting fibers are observed to bridge the larger fibers (B). At extremely high magnification the amorphous fibers appear to be made up of filaments containing systematically spaced striations (C). Models corresponding to the morphological features of the filaments have been proposed by Gotte (16). Three potential arrangements are given in D. The arrangements from left to right correspond to elastin in its stretched, relaxed and highly relaxed states. Figure E is our attempt to integrate the morphological features with chemical structure (14). The polymeric chains comprising elastin appear to be ordered in the form of β-spiral and α-helical segments. It is proposed that two polypeptide chains are cross-linked to comprise the filaments. The filaments in turn may be a reticulum of randomly cross-linked chains, if the bonding between the filaments arises from non-covalent forces (12). A possibility for the non-covalent cross-linking of the filaments is interactions involving calcium ions (14, 17, 18).

Table I

AMINO ACID COMPOSITION (EXPRESSED AS RESIDUES PER 1000 TOTAL
RESIDUES) OF TYPICAL MATURE ELASTIN, MATRIX COLLAGEN AND
MICROFIBRILLAR PROTEIN PREPARATIONS.

Amino Acid[a]		Crosslinked Elastin[b]	Bone Matrix Collagen[c]	Microfibrillar Protein[d]
Gly	(G)	332	338	110
Ala	(A)	228	112	65
Val	(V)	138	20	56
Pro	(P)	117	121	64
Hypro	(P)	16	107	--
Ile	(I)	25	12	48
Leu	(L)	60	23	69
Tyr	(Y)	6	2	36
Phe	(F)	29	13	38
Thr	(T)	10	14	56
Ser	(S)	10	39	62
Asp+Asn	(D+N)	8	36	114
Glu+Gln	(E+Q)	16	82	114
Met	(M)	--	6	16
1/2 Cys	(C)	--	--	48
His	(H)	--	5	15
Arg	(A)	5	39	45
Lys	(K)	3	27	45
Ides[e]		1	--	--
Des[e]		1.5	--	--
LNL[e]		1.2	N.C.	--
Hexose %		0	1	N.C.

[a]Common single letter abbreviations are given in parenthesis.
[b]Isolated from bovine ligamentum.   [c]Bovine bone matrix collagen.
The lysine-derived crosslinking amino acids are not calculated
(N.C.).   [d]Composition taken from ref. 2.   [e]See text and Figure
4.

Table II

COMMON AMINO ACID SEQUENCES IN NON-CROSSLINKED ELASTIN.

Repeating units[a]

Tetrapeptides: GGVPGAVPGGVPGGVFFPGAGLGGLG

Pentapeptides: YGAAGGLVPGAPGFG PGVGVPGVGVPGVGVPG(S)GV(P)GV(G)V
                  PGV(G)(V)

Hexapeptides: AAQFGLGPGIGVAPGVGVAPGV(G)VAPGVGV(A)PGVGVA
                  P(X)I

Examples of small tryptic peptides containing Ala- and Lys-rich
  sequences[b]

Sequence	Moles/mole protein
AAAK	6
AAK	6
SAK	2
APGK	2
AK	1
YGAK	2

---

[a]Amino acid sequences of specific tryptic peptides found in
porcine tropoelastin. When tentative assignments are given, the
designations are in parentheses. The common repeating units are
underlined. In certain instances liberty was taken in defining
a common repeating unit when there was only amino acid difference.
These sequences are common to the extensible regions in elastin.
The tetra, penta or hexa repeats appear to correspond to 15 to
25 percent of the total residues in the protein. The source for
the sequence data is reference 1.

[b]Sequences commonly found in the regions of elastin that are
eventually involved in crosslinking (cf. Figure 4).

cleavage of the non-crosslinked forms of elastin into discreet
subunits ranging in molecular weights from 12,000 to 70,000
(26,27). In addition the enzyme, lysyl oxidase, which is involved
in the crosslinking of elastin is also secreted (cf. refs. 28-35
and the section on Formation of Stable Elastin Fibrils).

The exact form in which non-crosslinked elastin is secreted
from smooth muscle cells is yet to be clearly defined. Foster et
al. (36) have suggested that a non-crosslinked elastin (proelastin) is secreted from smooth muscle cells in a form that is
approximately 120,000 to 140,000 daltons. They have suggested
that proelastin is cleaved to smaller molecular weight forms of
non-crosslinked elastin. It should be noted, however, that this
view is not entirely supported by data from other laboratories.
There are two reports on the use of isolated mRNA from chick aorta
suggesting only a 70,000 dalton non-crosslinked elastin is the
major product of translation (37,38). There is also a recent
report suggesting that aortic mRMA translates a 200,000 dalton
putative elastin product (39). We have recently isolated a non-
crosslinked elastin from the aortas of copper deficient chicks
that appears to be 100,000 daltons (27). Its amino acid composition is similar to that for tropoelastin (Table III). A major
problem in resolving these points is that the trypsin-like
proteinase associated with elastin is not easily denatured or
separated from the non-crosslinked forms of elastin. The
proteinase is also not readily inhibited by commonly used
inhibitors for trypsin-like proteinases (26).

In keeping with the concept of several forms of soluble
elastin, Figure 2 outlines the various steps which are envisioned
to be involved in the formation of the elastin fibril associated
with the microfibrillar components. The process, at least in
concept, is not entirely dissimilar to the processing and
synthesis of collagen fibrils (40). Once released from ribosomes,
it appears that non-crosslinked elastin is incorporated into
fibrils in a matter of minutes (27). Although it is not clear
what exact role the proteinase(s) plays, limited proteolysis could
act as signals for other post-translational events, such as cross-
linking. Alternately, proteolysis may control the net amounts of
elastic fibers synthesized during given periods of development.
Nevertheless, unique with regard to other examples where
proteinases play a role in protein regulation, elastolytic
proteinase(s) appears to function in normal development at early
steps in elastogenesis. It is of interest to note that, to date,
no true elastinase that readily degrades mature elastin in its
crosslinked state has been isolated from elastin-secreting cells.
Although the enzyme elastase has been studied extensively, one
should keep in mind that it has only been isolated from organs,
such as the pancreas, involved in digestive functions and
phagocytic cells, such as macrophages and leucocytes.

With respect to factors that cause stimulation of elastin
synthesis in tissues, there is some evidence to suggest that

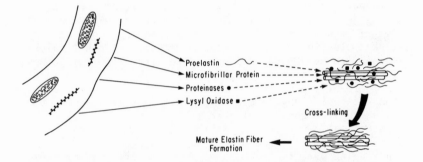

Figure 2.   Synthesis of mature elastin fibers. Some evidence suggests the possibility for proforms to elastin that appear as the first products of translation. These products are cleaved to tropoelastin (27), which appears to combine with microfibrillar protein. Although post-translational events important to the synthesis of the microfibrillar protein have not been defined, it is clear that it is a major component on which is organized or assembled the profibrillar forms of elastin. Cross-linking is catalyzed by lysyl oxidase, a copper-requiring protein (30). Recent information on the elastin proteinase(s) involved in tropoelastolysis would suggest that proteolysis may also play a role in elastin fiber formation (24).

Table III

AMINO ACID COMPOSITION OF TROPOELASTIN AND A PUTATIVE TROPO-
ELASTIN PRECURSOR (EXPRESSED AS MOLES PER 1000 MOLES OF AMINO
ACID RESIDUE).

Amino Acid	Tropoelastin Precursor	Tropoelastin
	(90,000-100,000 daltons)	(72,000 daltons)
Lys	40	42
His	trace	0
Arg	12	6
Hypro	8	8
Asp+Asn	15	5
Thr	17	10
Ser	13	8
Glu+Gln	23	13
Pro	121	127
Gly	315	335
Ala	165	175
1/2 Cys	0	0
Val	169	177
Met	0	0
Ile	16	18
Leu	44	54
Tyr	12	11
Phe	25	30

certain steroid hormones may alter net synthesis (41-43). Also,
there is evidence that suggests in certain tissues elastin
synthesis occurs in response to mechanical activity (44). Cells
that produce elastin when grown on preformed insoluble elastic
fibers will secrete greater quantities of matrix proteins if the
fibers are stretched and relaxed in culture than if they are
stationary or minced and agitated. In lung, new elastin synthesis
also follows the acute destruction of elastin caused by inhalation
of elastase or papain (45). None of these observations, however,
has been put on a firm basis at the molecular level.

Formation of Stable Elastin Fibrils

One of the most important steps in stabilizing elastin

fibrils is the formation of crosslinks. The crosslinks result
from the oxidation of specific lysyl residues. As mentioned above,
these residues are located within what appear to be defined cross-
linking regions in the polypeptide chains making up the fibrils
(28-35,46,47). The enzyme responsible for the oxidation is lysyl
oxidase. The mechanism of oxidation is probably similar to an
oxidative deamination (28,29). Lysyl oxidase requires copper
(30) and is inhibited by a family of lathyrogens, such as
β-aminopropionitrile (31). It has been purified from a variety
of connective tissue sources. However, there is still no clear
definition regarding its specificity towards specific substrates
(32,33). For example, elastin may serve as a substrate for lysyl
oxidases obtained from collagen-rich sources that do not contain
elastin (28-35).

It has been demonstrated by Rayton and Harris (30)
that the role of copper, in addition to its presumed role as a
cofactor, is related to the induction of lysyl oxidase.
Cycloheximide, but not actinomycin D, completely inhibits the
incorporation of $^{64}Cu^{II}$ into lysyl oxidase. They suggest that the
mechanism may be similar to the induction of ferritin by iron.
It is also of interest that when copper bound to serum proteins is
added to cultures of minced aorta obtained from copper-deficient
chicks, the amount of copper required for induction of lysyl
oxidase is one-tenth to one-twentieth of that required when
copper salts are added. Homogenizing the tissue or incubating
it under N_2 or in the cold blocks the appearance of the enzyme.

Under normal conditions it would appear that the enzyme is
secreted from cells in close association with its substrate. By
conventional extraction methods (physiological buffers), most of
the lysyl oxidase activity is not released from insoluble connective tissue fibers (33,34). It is only released after extraction with denaturants, such as urea. Further, it is difficult to
handle in solution because of its tendency to form aggregates.
Lysyl oxidase is rich in cysteine residues which may facilitate
formation of polymeric forms (33).

It has also been difficult to study the enzyme because
native substrates and inhibitors of the enzyme are extracted with
it into urea and thus must first be dissociated and removed before
an estimation of true activity can be obtained (44,45). Further,
there are no well-characterized substrates available for the
routine assay of lysyl oxidase. The standard assay is a procedure
described by Pinnell and Martin (31). The substrate is prepared
from embryonic chick aortas after culture in vitro in the presence
of 4,5- or 6-^{3}H-lysine and inhibitors of endogenous lysyl
oxidase (cf. Figure 3).

With such an ill-defined substrate, and the requirement of
a partial purification of lysyl oxidase before assay (46,47), the
examination of the enzyme's role in maturation or pathological
processes has been less than quantitative. However, the availability of the purified enzyme has allowed several investigators

to examine the interaction of lysyl oxidase with its substrates. It is now clear that the enzyme only acts on collagen and perhaps elastin when these proteins are in the form of fibrils (28,29,35,46). For example, only tropocollagen serves as a substrate, not dissociated subunits, such as collagen α-chains (35).

Following the oxidation of peptidyl lysine there is little evidence to date to suggest a role for other enzymes in crosslinking. The conversion of peptidyl lysine to peptidyl allysine (Figure 3) appears to set the stage for the spontaneous formation of crosslinks. The major requirements are probably most related to the conformational state of the protein, the location and juxaposition of lysyl derivatives, and what might be viewed as environmental factors, e.g., oxygen tension (28,29).

For purposes of this manuscript, we wish to concentrate only on the steps leading to the formation of desmosines, amino acids found predominantly in elastin. With respect to their formation, the following suggests their spontaneous formation from peptidyl lysine and the oxidation product, peptidyl allysine. Narayanan et al. (28,29) have shown that when purified lysyl oxidase and non-crosslined elastin, specifically tropoelastin, are incubated together, the desmosines are formed. Desmosine formation, however, only occurs at temperatures that favor fibrillar arrangements of tropoelastin. Subsequently, it is felt that the maturation of non-crosslinked elastin into cross-linked elastin appears to involve only two major steps, namely insolublization through the formation of fibrils and fixation of the fibrils by crosslinking.

To form the desmosines, three peptidyl allysine molecules and a molecule of peptidyl lysine must condense. The steps in condensation probably involve the formation of 1,2-dihydropyridines and 1,4-dihydropyridines as shown in Figure 4 (19-24,46,48). Several kinds of chemical evidence (46,48) suggest that the hydropyridines are easily oxidized under normal oxygen tension to corresponding pyridinium ions, such as the desmosines (isodesmosine or desmosine). The exact pathway by which the desmosines are formed, however, is still not clear.

Currently, there are at least two views related to the mechanism by which the desmosines are formed (19). These include the direct reaction of the so-called allysine aldol (cf. Figure 4) with dehydrolysinonorleucine to form desmosines, or alternatively, the reaction of dehydromerodesmosine with an allysine residue. The first mechanism would require the formation of the allysine aldol and dehydrolysinonorleucine as intramolecular crosslinks. The second mechanism would result from the stepwise addition of two allysines and lysine to form dehydromerodesmosine (via Michael additions) and then condensation with a fourth allysine residue. The major problem in resolving these points is the difficulty of sequencing around intra- and intermolecular crosslinks in a manner to provide definitive information.

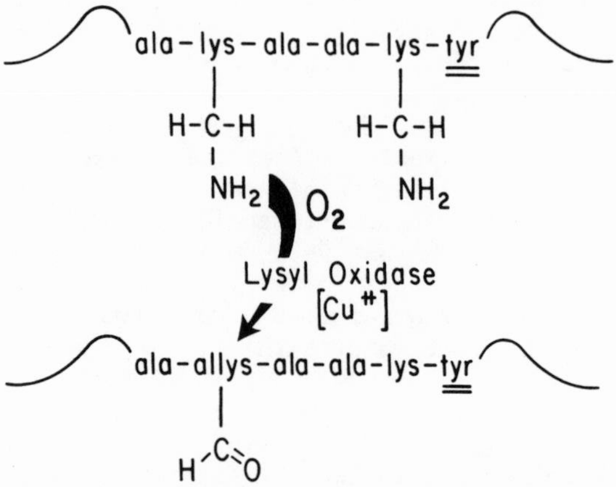

Figure 3. Lysyl oxidase. The enzyme, lysyl oxidase, appears to seek out lysyl resi-dues in alanyl- and lysyl-rich regions in the profibrillar forms of elastin. The presence of an aromatic amino acid residue adjacent to lysine appears to block its oxidation. The product of oxidation is peptidyl α-aminoadipic-δ-semialdehyde. Assays for the enzyme against elastin involve first the preparation of an elastin-rich pellet containing ^{3}H-lysyl residues labeled in the 6 or 4,5 position. This is usually accomplished by incubating embryonic chick aortas in medium containing ^{3}H-lysine plus β-aminopropionitrile (BAPN) to inhibit endogenous lysyl oxidase activity. BAPN is then removed leaving behind an elastin-rich residue in which the profibrillar forms of elastin labelled with ^{3}H-lysine are only partially crosslinked. When lysyl oxidase preparations are added to this residue the release of tritium represents the assay for activity. It has also been demonstrated that tropoelastin, when incubated with lysyl oxidase, forms α-aminoadipic-δ-semialdehyde and eventually crosslinks as shown in Figure 4.

There is still no way of determining whether or not a given desmosine crosslinks 1, 2, 3, or 4 polypeptide chains of elastin. Based on model studies, however, the most favorable arrangement would be expected if only two chains are crosslinked together by a desmosine (19). This extends from observations that polyalanyl-rich peptides typically favor α-helical conformations and that it is difficult to interconnect more than two polypeptide chains around any given desmosine. With regard to the other amino acids that could potentialy crosslink elastin, the exact number of dehydrolysinonorleucine, dehydromerodesmosine and allysine aldol residues that are involved as intra- or intermolecular crosslinks, and the extent to which these residues may be reduced to form stable crosslinks is not known.

The Elastic Fiber; Alterations During Maturation

With the above overview of elastic fiber formation, attention may now be directed at changes which occur in elastin upon maturation of the fiber. As mentioned above, the fact that mature elastin and other components of the elastic fiber are insoluble after crosslinking dictates that harse procedures have to be used in order to isolate the protein. Because of this, it is often difficult to determine whether or not one is dealing with a pure elastin or a mixture of other structural protein components and elastin. For example, when elastins as defined by alkali insolubility are isolated from matrix synthesized by cultures of smooth muscle cells, it has been observed that the composition of this material is altered upon maturation (49). Whether the changes in composition represent changes that are due to differences in elastin or merely indicate differing amounts of other structural proteins behaving as elastin during isolation has been difficult to clarify. Obviously differences in composition and the rather harsh isolation procedures raise serious questions when one is assessing homogeneity.

Further, when one examines crosslinks in aged tissues using harse techniques, it is also necessary to ask whether or not any of the changes observed are due to age-related events or the method employed. An example of the latter is reported by Barnes et al. (50). They examined elastin from guinea pig aorta that had been previously radiochemically labeled with ^{14}C-lysine. When this product (obtained after alkali-extraction of the aortas) was further treated with boiling oxalic acid in order to obtain a soluble, but crosslinked product, "α-elastin", the radioactivity associated with the desmosines in the solubilized elastin was greater than that in the original starting material. The reason for this is probably related to the conditions used in the isolation, which forced the formation of desmosine from its precursors in a manner in keeping with the scheme shown in Figure 4.

With respect to data on differences in the crosslinking amino

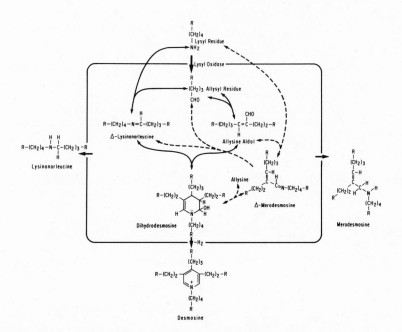

Figure 4. A general scheme for the formation of desmosine from lysine. The figure was adapted from W. R. Gray (19). Reactions within the box are considered to be reversible. Steps that result in the formation of lysinonorleucine, merodesmosine or desmosine are considered irreversible. Two separate pathways by which desmosine may be formed are indicated by the solid or broken lines.

acid content of elastin upon maturation, one of the best examples
is work reported by Francis et al. (51). Elastins were purified
by various methods from the ligamentum of cattle at different ages
and were analyzed for lysine-derived crosslinks (Table IV). In
fully mature elastins from adults, the main crosslinks were
desmosine, isodesmosine, lysinonorleucine, and the allysine aldol
product. Trace amounts of merodesmosine were also found. During
the normal maturation of the elastic fiber, the amounts of
desmosine, isodesmosine and lysinonorleucine were increased;
whereas the aldol condensation product and lysine residues were
decreased. Elastins from young animals contained significantly
higher amounts of dehydromerodesmosine and dehydrolysinonorleucine
than from adult animals. These differences in distribution
probably reflect the difference in the distribution of "young"
versus "old" elastin in elastic fibers. Also, it should be
noted that only 50% of the lysine in the profibrillar forms of
elastin is accounted for as crosslinks. Obviously the question of
elastin crosslinking with respect to aging may only be resolved
when the other forms of presumably lysine-derived crosslinks are
firmly established (compare the values for lysine in Table III
with those in Table IV).

When the metabolic turnover of elastin in arterial tissue or
in lung is examined, it is extremely difficult to demonstrate
active turnover. Once an elastin fiber is formed it appears to
be fixed. The turnover of rat aorta elastin is best measured in
years (8). Data shown in Figure 5 also suggests negliable turnover. The animal used for this study, the Japanese quail, was
chosen because it fully matures at 5-6 weeks of age. Similar to
the rat its elastin appears to turn over in amounts best estimated
in years.

Other than directly measuring turnover, it has also been the
convention of some to use the ratio of lysine to desmosine or
other crosslinks as an index of maturation (8). In certain
instances this may be justified. However, if only subtle changes
occur, it is again very difficult to distinguish between artifacts
and a process that signifies the formation of newly synthesized
elastic fibers or defects in crosslinking. Where the evidence is
clear comes from the studies on nutritional copper deficiency
lathryism, and genetic mutants with crosslinking defects (cf. ref.
46 and references cited). With these conditions death often
results from pathologies associated with the decrease in crosslinks. However, in other chronic disease conditions, such as
emphysema and atherosclerosis, where elastin metabolism is an
important component or consideration, the data are not clear.
For example, there are several reports that indicate many of the
previously reported changes in elastin composition during
atherosclerosis are related to the methods used in the initial
isolation of the elastins (52,53). Although in certain forms of
emphysema there appears to be evidence that elastic fibers are
destroyed, the data related to compositional changes and alter-

TABLE IV

CONCENTRATIONS OF CROSSLINKS AND LYSINE RESIDUES IN ELASTIN ISOLATED FROM THE LIGAMENTUM NUCHAE OF CATTLE OF INCREASING AGE[a]

Isolation method	Fetal			1 week old			3 years old			12 years old		
	Alkali	Autoclaved	Formic acid	Alkali	Autoclaved	Formic acid	Alkali	Autoclaved	Formic acid	Alkali	Autoclaved	Formic acid
Isodesmosine	0.73	0.82	0.75	1.06	1.09	0.89	1.26	1.31	1.28	1.16	1.18	1.19
Desmosine	0.79	1.01	0.81	1.37	1.30	1.15	1.74	1.73	1.75	1.70	1.75	1.76
Dehydromerodesmosine	0.76	0.62	0.71	0.91	0.79	0.81	0.04	0.03	0.10	0	0	0
Merodesmosine	0.18	0.17	0.18	0.22	0.26	0.23	0.22	0.22	0.19	0.21	0.23	0.24
Dehydrolysinonorleucine	0.22	0.13	0.15	0.10	0.07	0.25	0.04	0.03	0.02	0.08	0	0
Lysinonorleucine	0.78	0.66	0.50	1.14	0.83	0.61	0.87	0.85	0.85	0.86	0.82	0.80
Allysine	3.44	3.43	3.20	2.92	2.76	2.93	2.15	1.86	1.95	1.70	1.60	1.47
lysine	9.10	10.80	9.00	6.10	7.00	5.90	3.50	3.80	3.50	2.30	2.60	3.00
Total lysine	26.88	28.93	25.61	27.53	27.03	24.76	22.40	22.19	22.13	19.65	19.85	20.06

[a]Concentrations are in residues/1000 amino acid residues. When cross-links are expressed as lysine equivalents, desmosine and isodesmosine each equal four, dehydromerodesmosine and merodesmosine each equal three, and dehydrolysinonorleucine, lysinonorleucine, and the aldol-condensation product each equal two lysine residues. Data taken from Francis et al. (51).

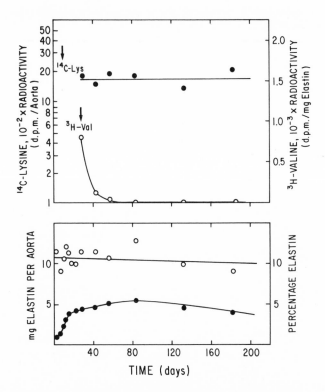

Figure 5. Turnover of arterial elastin. Mature elastin was isolated from the aortas of Japanese quail over a 200 day period. The top figure indicates the turnover of ^{14}C-lysine incorporated into elastin. The birds were injected with ^{14}C-lysine at 9 days post-hatching and the first elastin samples were isolated at 4 weeks in order to start at a point where recycling of ^{14}C-lysine from the degradation of other proteins would be minimized. The values represent the total radioactivity per whole organ so that growth would not compromise the estimates. Also shown is the specific activity of injected ^{3}H-valine per milligram of elastin. These values were obtained 24 hours after injection in order to roughly estimate the potential for new elastin synthesis. There was little evidence for new synthesis following 8 weeks of development at a point where the birds are sexually mature. The bottom figure shows expressions of elastin as the total per aorta (●—●). Note the increase during the early stages of development reflecting the growth of the aorta. Expressed as a percentage (○—○), the elastin content was approximately 10–12% of the weight of the aorta throughout the period examined. The data indicate the lack of turnover of the protein. In fact, no true estimate of $t_{1/2}$ could be obtained from the ^{14}C-lysine curve.

ation in crosslink content may only reflect some attempt by the tissue to replace the degraded elastin by new elastic fibers (45).

Although outside the scope of this review, we would be remiss not to mention that there is an abundance of morphological evidence that suggest significant alterations in diseased or aged elastic fibers (54). Often old fibers appear frayed and nicked. Because of the three dimentional network of elastic fibers, it is possible to effect considerable damage without altering the bulk of the protein comprising the fibers. This may be one reason why elastolytic processes result in changes that are not easily detected in vivo using biochemical techniques.

Also, our comments would not be complete without mentioning work by Kagan and his co-workers dealing with the general phenomena of elastolysis (55-58). Because of this work it is now generally recognized that elastin is more easily hydrolyzed if it is presented to elastolytic enzymes, specifically elastase, in the form of a lipid-protein complex. The most suitable lipids appear to be long-chain fatty acids or their analogs (55,56). For example, preincubation of mature elastin with linoleic acid increases its suspectibility to elastolysis by elastase by an order of magnitude greater than delipidated elastin (55). Elastase prefers elastin as substrate if it is anionic in character, which is the case when anionic detergents or fatty acids are bound to elastin. Cationic detergents do not stimulate elastolysis. From the standpoint of elastin turnover in vivo a property of the protein which may provide protection from proteolytic enzymes resembling elastase is its cationic character (57,58). In fact, a significant feature of the protein that may protect against normal elastolysis is the observation that over 70% of the glutamyl and aspartyl residues in the protein appear to be amidated (58).

Concluding Remarks

Elastin is a very novel protein and elastomer. Its alternating regions rich in apolar amino acids and alanine and lysine-derived crosslinks distinguishes it from collagen and other structural proteins. Its anisotropic structural characteristics provides the structural chemist with an interesting model to study elasticity. Its long biological half-life is unique. Its resistance to harse reagents and many proteinases also distinquishes elastin from most proteins.

Lastly, some major diseases which inflict man are associated with elastin metabolism. Two of the most important are emphysema and athersclerosis in which alterations in the elastic fibers appear to play a role. Obviously, experiments related to the understanding of disease processes involving elastin need to take into account the extent to which these processes may be reversed.

Acknowledgments

Mike Lefevre is funded from a training grant from the National Institute of Dental Research (DE-07001). Preparation of this manuscript was supported in part from NIH grants HL-18918 and HL-15965.

LITERATURE CITED

1. Sandberg, L. B.   Int. Rev. Connect. Res., 1976, 7, 159.
2. Ross, R.   J. Histochem. Cytochem., 1973, 21, 199.
3. Jackson, D. S.; Cleary, E. G.   Methods Biochem. Anal., 1967, 15, 25.
4. Ross, R.; Bornstein, P.   J. Cell Biol., 1969, 40, 366.
5. Richmond, V.   Biochim. Biophys. Acta, 1974, 351, 173.
6. Serafini-Fracassini, A.; Fielo, J. M.; Rodger, G. W.; Spina, M. Biochim. Biophys. Acta, 1975, 386, 80.
7. Rasmussen, B. L.; Bruenger, E.; Sandberg, L. B.   Anal. Biochem., 1975, 64, 255.
8. Rucker, R. B.; Tinker, D.   Int. Rev. Exp. Path., 1977, 17, 1.
9. Gosline, J. M.   Int. Rev. Connect. Res., 1976, 7, 211.
10. Harkness, R. D.   In "Treatise on Collagen" (B. S. Gould, ed.) Academic Press: New York 1968; p. 247.
11. Partridge, S. M.   Advan. Protein Chem., 1962, 17, 227.
12. Serafini-Fracassini, A.; Field, J. M.; Hinnie, J. J. Ultrastructure Res., 1978, 65, 190.
13. Cleary, E. G.; Cliff, W. J.   Expl. Mol. Path., 1978, 28, 227.
14. Urry, D. W.   Advan. Exp. Med. Biol., 1975, 43, 211.
15. Hoeye, C. A.; Floey, P. J.   Biopolymers, 1974, 13, 677.
16. Gotte, L.   Advan. Exp. Med. Biol., 1977, 79, 105.
17. Urry, D. W.; Krivacia, J. R.; Haider, J.   Biochem. Biophys. Res. Commun., 1971, 43, 6.
18. Rucker, R. B.; Ford, D.; Riemann, W.; Tom, K.   Connect. Tiss. Res. 1974, 14, 317.
19. Gray, W. R.   Adv. Expl. Med. Biol., 1977, 79, 285.
20. Foster, J. A.; Rubin, L.; Kagan, H. M.; Franzblau, C. J. Biol. Chem., 1974, 249, 6191.
21. Gerber, G. E.; Anwar, R. A.   J. Biol. Chem., 1974, 249, 5200.
22. Gerber, G. E.; Anwar, R. A.   Biochem. J., 1975, 149, 685.
23. Gray, W. R.; Sandberg, L. B.; Foster, J. A.   Nature (London), 1973, 246, 461.
24. Mecham, R. P.; Foster, J. A.   Biochem. J., 1978, 173, 617.
25. Keith, D. A.; Paz, M. A.; Gallop, P. M.; Glimcher, M. J. J. Histochem., 1977, 25, 1154.
26. Mecham, R. P.; Foster, J. A.   Biochemistry, 1977, 16, 3825.
27. Heng-Khoo, C. S.; Rucker, R. B.; Buckingham, K. W.   Biochem. J., 1979, 177, 559.
28. Narayanan, A. S.; Page, R. C.   J. Biol. Chem., 1976, 251, 1125.

29. Narayanan, A. S.; Page, R. C.; Kuzan, F.; Cooper, C. G.
    Biochem. J., 1978, 173, 857.
30. Rayton, J. K.; Harris, E. D. J. Biol. Chem., 1979, 254, 621.
31. Pinnell, S. R.; Martin, G. R. Proc. Natl. Acad. Sci. U.S.A.,
    1968, 61, 708.
32. Siegel, R. C.; Fu, J. C. C. J. Biol. Chem., 1976, 251, 5779.
33. Stassen, F. L. H. Biochim. Biophys. Acta, 1976, 438, 49.
34. Narayanan, A. S.; Siegel, R. C.; Martin, G. R.
    Arch. Biochem. Biophys., 1974, 162, 231.
35. Siegel, R. C. Proc. Natl. Acad. Sci. U.S.A., 1974, 71, 4826.
36. Foster, J. A.; Mecham, R. P.; Rich, C. B.; Cronin, M. F.;
    Levine, A.; Imberman, M.; Salcedo, L. I. J. Biol. Chem.,
    1978, 253, 2797.
37. Burnett, W.; Rosenbloom, J. Biochem. Biophys. Res. Commun.,
    1979, 86, 478.
38. Ryhanen, L.; Graves, P. N.; Bressan, G. M.; Prockop, D. J.
    Arch. Biochem. Biophys., 1978, 185, 344.
39. Krawetz, S. A.; Anwar, R. A. Fed. Proc., 1979, 38, 300.
40. Bornstein, P. Annu. Rev. Biochem., 1974, 43, 567.
41. Wolinsky, H. Cir. Res., 1972, 30, 341.
42. Wolinsky, H. J. Clin. Invest., 1972, 51, 2552.
43. Wolinsky, H. Cir. Res., 1973, 32, 543.
44. Leung, D. Y. M.; Glagov, S.; Mathews, M. B. Science, 1976,
    191, 475.
45. Goldstein, R. A.; Starcher, B. C. J. Clin. Invest., 1978,
    56, 1286.
46. Rucker, R. B.; Murray, J. Am. J. Clin. Nutr., 1978, 31, 1221.
47. Rowe, D. W.; McGoodwin, E. B.; Martin, G. R.; Grahn, D.
    J. Biol. Chem., 1977, 252, 939.
48. Davis, N. R. Biochim. Biophys. Acta, 1978, 538, 258.
49. Faris, B.; Salcedo, L. L.; Cook, V.; Johnson, L.; Foster,
    J. A.; Franzblau, C. Biochim. Biophys. Acta, 1976, 418, 93.
50. Barnes, M. J.; Constable, B. J.; Kodicek, E. Biochem. J.,
    1969, 113, 387.
51. Francis, G.; John, R.; Thomas, J. Biochem. J., 1973, 136, 45.
52. Keeley, F. W.; Partridge, S. M. Atherosclerosis, 1974, 19,
    287.
53. Spina, M.; Garbin, G. Atherosclerosis, 1976, 24, 267.
54. Bader, L. Pathology, 1973, 5, 269.
55. Jordan, R. E.; Hewitt, M.; Kagan, H.; Franzblau, C.
    Biochemistry, 1974, 13, 3497.
56. Kagan, H. M.; Crombie, G. D.; Jordan, R. E.; Lewis, W.;
    Franzblau, C. Biochemistry, 1972, 11, 3412.
57. Kagan, H. M.; Milbury, P. E.; Kramsch, D. M. Cir. Res., 1979,
    44, 95.
58. Kagan, H. M.; Jordan, R. E.; Lerch, R. M.; Mukherjee, D. P.;
    Stone, P.; Franzblau, C. Adv. Exp. Biol. Med., 1977, 79, 189.

RECEIVED October 18, 1979.

Photooxidative Damage to Mammalian Cells and Proteins by Visible Light

L. PACKER and E. W. KELLOGG, III

Membrane Bioenergetics Group, Lawrence Berkeley Laboratory,
University of California, Berkeley, CA 94720

During the last few decades mechanisms for the initiation
and propagation of cellular damage by ultraviolet and ionizing
radiation have received special attention. Photooxidative damage
by visible irradiation ($\geq$ 400 nm) has received less attention in
biological systems. Most investigations have focused on dye
photosensitized reactions, rather than on chromophores found in
situ, such as flavins and hemes, which normally act as cofactors
in biological oxidation-reduction reactions. The visible light
system can serve as an amplified model portraying oxidative stress
in aerobic cells in that pro-oxidant substances (O_2^-, $^1O_2$, $\cdot OH$,
H_2O_2, etc.) produced during normal metabolism, are easily generated under photooxidative stress. However, in itself the effect
of visible light on biological systems has marked relevance in
that it is a factor to which almost all organisms are exposed and
must contend. In the present article we will review studies carried out in our laboratory on the effects of visible irradiation
and O_2 in a variety of target systems ranging from cultured mammalian cells to purified catalase. We will relate these studies
of photooxidative damage to a scheme for the propagation of intracellular damage (Fig. 1) which traces a number of the possible
pro-oxidant and anti-oxidant pathways found in the cell.

Prooxidative reaction pathways. For visible light to affect
cellular components it must first be absorbed. Hemes with a
$\lambda_{max} \simeq 450$, $\varepsilon = 28,750$ $M^{-1}cm^{-1}$ (for catalase heme) ([1]) and flavins
with a $\lambda_{max} \simeq 445$, $\varepsilon = 12,500$ $M^{-1}cm^{-1}$ (for FMN) ([2]) are the most
probable sites of visible light absorption and oxygen activation.
The excited sensitizer can chemically react directly with other
compounds by a Type I process ([3]); Eqn 1.

$$\text{Sens} \xrightarrow{\text{hv}} {}^1\text{Sens} \longrightarrow {}^3\text{Sens} \xrightarrow{\text{substrate}} \text{H or } e^- \text{ transfer} \qquad (1)$$

Inasmuch as these photosensitizers serve as enzymatic cofactors,
their excitation and reaction at the active site could lead to
enzyme inactivation.

However, in the presence of oxygen, Type II reactions are likely to occur producing $^1O_2$ and O_2^- (Eqns 2 and 3):

$$^3Sens \xrightarrow{O_2} {}^1O_2 + Sens. \tag{2}$$

$$^3Sens \xrightarrow{O_2} O_2^- + Sens^+ \tag{3}$$

Both flavins and hemes have been shown to participate in Type I and II reactions.

As illustrated in Fig. 1, reactions of nonprotein and protein-bound coenzymes and metals with O_2 or with H_2O_2 can serve as routes of production of damaging oxygen radical species. H_2O_2 can be formed from the dismutation of O_2^- (Eqn 4):

$$2O_2^- + 2H^+ \longrightarrow H_2O_2 + O_2 \tag{4}$$

which can then form $\cdot OH$ by a one electron reduction process. For example, Equations 5 and 6

$$O_2^- + H_2O_2 \longrightarrow \cdot OH + OH^- + O_2 \tag{5}$$

$$Fe^{+2} + H_2O_2 \longrightarrow \cdot OH + OH^- + Fe^{+3} \tag{6}$$

give reactions that have considerable experimental support (4), although the interaction of O_2^- with H_2O_2 may be Fe^{+3} mediated

$(Fe^{+3} + O_2^- \rightleftharpoons O_2 + Fe^{+2})$. Other evidence supports the idea

that O_2^- and H_2O_2 can give rise to $^1O_2$ as well as $\cdot OH$ by a reaction similar to equation 5 (5-7); however the actual mechanism of $^1O_2$ generation is unknown.

Hydroxyl radical ($\cdot OH$), with a reaction rate of k $\underset{\sim}{\sim}$ $10^9 M^{-1} sec^{-1}$ for most organic compounds, is probably the most reactive and damaging species found in biological systems. Singlet oxygen, although more selectively reactive than $\cdot OH$, reacts rapidly with compounds with amine groups or double bonds such as unsaturated fatty acids, amino acids, and nucleic acids. Some of the possible pro-oxidant pathways beginning with O_2, O_2^- and $\cdot OH$ are depicted in Fig. 1; these represent the most probable pathways of oxidative damage in mitochondria.

Antioxidative reaction pathways. The prevention of damage in cellular systems can be considered a two-level process. First, the cell would minimize the production and availability of prooxidant factors and substances. The compartmentalization of prooxidant enzymes in organelles such as mitochondria, and the sequestering of transition metals by specific proteins are examples of this level of defense. The second level of defense involves the scavenging and neutralization of pro-oxidants. These antioxidant pathways in mitochondria are depicted by darker lines in

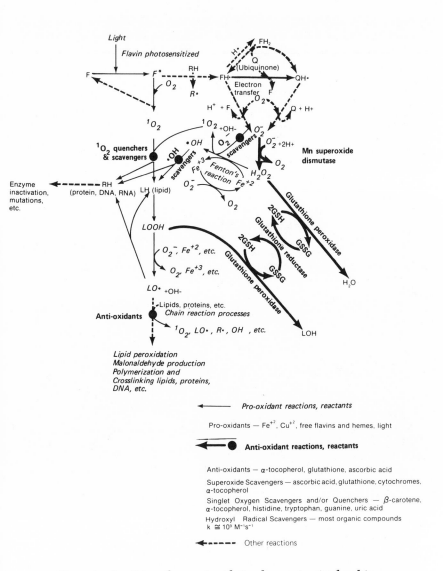

Figure 1.   *Possible pathways of oxidative damage in mitochondria*

Fig. 1, and show the quenching of $^1O_2$ by vitamin E, the dismutation of O_2^- by superoxide dismutase (SOD), the conversion of H_2O_2 to $H_2O_2 + O_2$ by glutathione peroxidase, and the scavenging of free radicals by antioxidants such as ascorbic acid in the aqueous phase and α-tocopherol in the lipid phase. These two antioxidants can work synergistically in preventing membrane damage (8). We will attempt to relate these proposed pathways of cellular damage to the patterns of damage actually found in our visible light studies.

Results

Cultured mammalian cells and isolated hepatocyte studies. There have been several recent reports of the damaging effects of visible light exposure (>400 nm) on various microorganisms (9,10) and cultured mammalian cells (11-13). We have reported that human diploid cells on exposure to visible light (14) and oxygen (>10%) (15) lost the ability to proliferate, while ultrastructural studies showed the presence of numerous damaged mitochondria in the illuminated cells (16). WI-38 human fibroblasts show a gradual decline in growth rate with exposure to visible light with younger cells (14) being more susceptible to photokilling, with partial protection observed on the addition of dl-α-tocopherol (vitamin E).

Studies with isolated hepatocytes (17) have led us to a characterization of the pattern of intracellular damage. Exposure of rat hepatocytes to visible light (400-720 nm) of intensity 300 mW/cm^2 over a 12 hr period results in a selective pattern of subcellular damage (Fig. 2). Virtually no release of lactate dehydrogenase or uptake of trypan blue was observed. The plasma membrane enzymes 5'-nucleotidase and β-leucyl naphthylamidase were only slightly inactivated. The plasma membrane thus appears highly resistant to damage.

Under the same conditions, however, other intracellular enzymes were markedly inactivated. Mitochondrial damage was indicated by a decrease in latency of chtochrome c oxidase and destruction of various enzyme activities in the following order: succinic dehydrogenase > succinate oxidase > glutathione peroxidase > NADH-cytochrome c oxidase > cytochrome c oxidase (Fig. 2). This pattern of inactivation is similar to the one found upon light exposure of isolated mitochondria (18), suggesting that continued studies with the in vitro system are indeed warranted. Lysosomal damage was also extensive, as indicated by the loss of latency and activity in the enzymes cathepsin c, acid phosphatase and N-acetyl-β-glucosaminidase. Some evidence of damage to microsomal membranes was indicated by a decline in glucose-6-phosphatase activity. The most light-sensitive enzyme was found to be catalase, an enzyme associated with the peroxisomal fraction. Another peroxisomal enzyme, urate oxidase, was relatively less suspectible to light damage. It is interesting to note that two

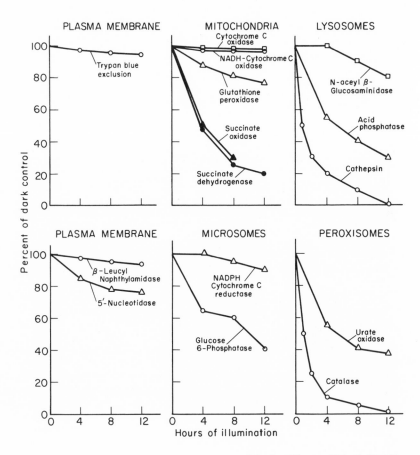

Figure 2. Enzyme photoinactivation in isolated hepatocytes (17). Hepatocytes in 0.25M sucrose at 8–10°C were illuminated in a shaking water bath at 300 mW/ cm² visible light (400–720 nm).

of the most light sensitive enzymes, catalase and succinate dehydrogenase, contain a heme and a flavin moiety, respectively, at
their active sites. Although further studies demonstrated the
oxygen dependence of their inactivation, scavengers of $^1O_2$, O_2^-,
and ·OH failed to protect, indicating that damage occurs at the
active site itself. Significantly, complete protection was af-
forded by substrates in both cases (18,19). Inactivation of other
enzymes lacking photosensitive cofactors presumably occurs by more
indirect reaction pathways. In addition to inactivation of enzymes, destruction of membrane lipids was indicated by lipid peroxidation measurements.

     Attempts to prevent visible light damage of succinate de-
hydrogenase showed that the addition of succinate + KCN was maxi-
mally effective. EDTA was effective in preventing succinate dehydrogenase inactivation and lipid peroxidation as measured by the
TBA test. It is well known that succinate + KCN and EDTA both can
act as reductants. Reducing conditions may protect against visible
light damage by reducing flavins, which cannot act as efficient
photosensitizers since they absorb very little visible light (ε =
870 $\underline{M}^{-1}cm^{-1}$ for FMN at 445 nm (2)). dl-α-Tocopherol and butylated
hydroxytoluene appeared effective against lipid damage but only
the latter antioxidant appreciably affected the pattern of enzyme
(succinate dehydrogenase) inactivation. Hepatocytes isolated from
rats fed with a vitamin E-deficient diet showed a marked increase
in susceptibility to lipid peroxidation compared to rats fed a
vitamin E-supplemented diet.

     Isolated mitochondria. Chance et al. (20) have shown that the
absorption spectrum of whole cells is qualitatively similar to
that of isolated mitochondria. Thus, visible light absorption by
cells may involve mitochondrial flavins or hemes as endogenous
photosensitizers. In isolated mitochondria we found that the inner energy transducing membrane can be extensively photooxidative-
ly damaged. Such bioenergetic parameters as maintenance of proton
and electrical potential gradients associated with coupling to ATP
synthesis were progressively inhibited following exposure of isolated mitochondria to light, whereas samples kept under identical
conditions for 12 hr in the dark showed no such changes (18).
These bioenergetic parameters show an interesting pattern of
change. Almost immediately after light exposure, uncoupling is
detected as shown by a stimulation of respiration, loss of ATP
synthesis, and increased ATP hydrolysis, all of which indicate uncoupling of electron transport from energetic gradients. Membrane
potential changes seem largely dependent upon protein inactivation
and occur at an earlier time period than the surface potential
changes. Membrane potentials are unaffected by lipid-soluble
antioxidants, whereas the surface potential changes occur at a
later time exposure and are partially reversed by antioxidants,
indicating that membrane lipids contribute to the surface electrical potential when measured with amphipathic spin labeled probes.

Electron transport components examined by spectrophotometric and EPR methods identified the major target of photooxidative attack as the flavin dehydrogenases (18). These complexes contain in addition to flavins, iron sulfur (FeS) clusters and active SH groups as important functional components and these also showed evidence of destruction. Quinones, some of which are bound to dehydrogenase complexes or which exist as a quinone pool, also showed photoinactivation. However, all of the heme-containing cytochromes of the bc complex and of the cytochrome oxidase complex, and cytochrome c showed no inactivation despite their visible light absorption. Hence, flavin-, FeS- and quinone-mediated photooxidative processes appear involved in initiation and propagation of damage.

Experiments with water-soluble spin labels added to mitochondrial inner membranes have demonstrated that photodestruction (not reduction) of spin signal occurs with an action spectrum coinciding with flavins. This suggests that some of the membrane protein-bound flavin coenzyme is released following light/O_2 exposure, and that released flavin radicals can be detected in solution. Released flavins could initiate photosensitized reactions that would accelerate the photoinactivation process. Indeed, our previous studies (14) indicate that maximum photokilling of WI-38 cells occurred by illumination in the wavelength region of maximum absorption by flavins. This evidence supports the idea that the cytochromes, despite their possession of heme groups, do not mediate visible light damage by acting as photosensitizers. Propagation of damage likely involves radicals in both the lipid and aqueous phases. Evidence of peroxidized lipids can readily be discerned, but this can largely be prevented by adding membrane-soluble antioxidants such as vitamin E or butylated hydroxytoluene which apparently prevent lipid peroxidation but leave the pattern of enzyme damage largely unaffected. Other studies with submitochondrial particles demonstrated an oxygen dependence for photoinactivation of all mitochondrial enzymes tested (18).

Catalase photoinactivation. Studies demonstrated that the inactivation of catalase is oxygen dependent and can be prevented by substrates (100 μM methanol or ethanol), while antioxygenic substances in general (sucrose for ·OH, histidine for $^1O_2$, and 10 μg/ml superoxide dismutase for O_2^-) have little protective effect (see Table I) (19). Superoxide dismutase does, however, partially protect purified catalase added to the mitochondrial fraction, indicating that O_2^- produced during photooxidation of the mitochondrial fraction can inactivate catalase, probably by converting the active Compound I form to the inactive Compound II. Since catalase is a key enzyme in H_2O_2 metabolism, the importance of its inactivation both in in vivo and in vitro to the overall metabolic protective capacity of cells needs to be carefully characterized to identify its significance in the time sequence of damaging events. It is interesting to note that light with a wave-

Table I. Photoinactivation of Mitochondrial Fraction Catalase.
          Specificity of Protection[a]

Conditions	Activity (per cent)	
	Dark	Light
O time	100.0 ± 0.4	100.0 ± 0.4
Complete system	89.3 ± 1.4	23.6 ± 2.5
+ Superoxide dismutase (10 μg/ml)	85.7 ± 1.3	22.4 ± 0.5
+ Histidine (1 mM)	98.7 ± 0.6	31.4 ± 2.1
+ Ethanol (100 μM)	104.2 ± 9.4	96.6 ± 2.7

[a] The complete incubation system contained isolated mitochondria resuspended in 0.25 M sucrose to a final concentration
of 0.5 mg/ml protein. Light samples were exposed to an incan-
descent light source (a bank of 50 watt G.E. reflector bulbs)
with an intensity of about 15 mW/cm^2 for 2 hr. Six ml samples
were incubated in a slowly shaking water bath at 34° C; small
aliquots were removed for catalase assays. Samples run in
duplicate.

length responsible for maximal catalase inactivation, has been
implicated as having a major role in photooxidative damage of
cultured cells (21).

Discussion

Photooxidative damage pathways. Based upon the results with
cells and mitochondria, it is possible to construct a scheme to
account for the various possible pathways of initiation and propagation of damage to lipids and proteins by photooxidative processes. Our results suggest a definite order of events in biological photooxidation processes. First, one sees a rapid inactivation of enzymes containing light sensitive cofactors at the active site, a process that is oxygen dependent and apparently not
susceptible to inhibition by exogenously added scavengers of active oxygen species but which can be totally prevented by the
addition of substrates. Examples of such enzymes are succinic dehydrogenase (FAD), NADH dehydrogenase (FMN) and catalase (heme).
As their inactivation requires oxygen this implies that a type II
process occurs at the photosensitive cofactor, which produces an
active form of oxygen ($^1O_2$, O_2^-, etc.) which reacts with a susceptible group at the active site, causing damage and loss of activity. The production of such oxidative species by bound photosensitizer would obviously have much less effect on areas distal
from the active site. However, photooxidative damage eventually
can cause the release of the photosensitizer group, which could
then cause a far more generalized pattern of damage in the cell.
The release of free photosensitizers, such as flavins, would
be expected to act as $^1O_2$ generators (Eqn 2). Also, autooxida-
tion of flavins and certainly quinones generate O_2^- in mitochondria (as in Eqn 3). The degree to which the "Haber-Weiss reaction" (Eqn 5) and Fentòn reaction (Eqn 6) occur in vivo is still
uncertain. Currently, experiments are underway in several laboratories to obtain quantitative information on •OH radical generation by Equations 5 and 6 using, in particular, spin trapping methods. Thus, the characteristic •OH radical adduct of the spin
trap DMPO (5,5-dimethyl-1-pyrroline-N-oxide) can be shown to occur in mitochondria exposed to visible light (24) but it is still
unclear from which stage in the propagation of damage that these
•OH radicals arise.
In mitochondria our data indicate that a substantial release
of free flavins does occur (18). It would be primarily from such
free photosensitizers that damage to enzymes without photoactive
groups, and lipid peroxidation, would occur. The propagation of
damage through the initiation of oxidized lipid peroxides and
alkoxy radicals is also an area in which quantitative information
is required. Methods are now becoming available that use artificial lipid vesicles (22) and monolayer systems (23) to investi-
gate the rate, extent and nature of free radical mediated oxidative reactions in lipids which can determine how damage spreads

in the vertical and lateral modes through membranes. Propagation of cellular damage could occur in both the cytosol and lipid phases of the cell, acting synergistically in causing the total damage profile. Figure 1 indicates the multifarious damage processes and interactions that may occur after the release of free photosensitizers. A pattern for the chemical defense against photooxidative damage can be recognized, and knowledge of pro-oxidant and antioxidant pathways may help in devising nutritional means which could afford increased protection against oxidative damage. At the present time, however, it is clear that many unanswered questions remain as to both the existence and importance of the many possible oxidative mechanisms of biological damage processes in vivo.

Acknowledgment

Research reported in this article was supported by the Energy and Environment Division of the U. S. Department of Energy under contract No. W-7405-ENG-48 and Hoffmann-La Roche Inc.

References

1. Torii, K.; Ogura, Y., J. Biochem. (Tokyo), 1968, 64, 171.
2. Data for Biochemical Research, 2nd Edition (ed. by R. M. C. Dawson; D. C. Elliott; W. H. Elliott; K. M. Jones) 200, Oxford University Press, London, 1969.
3. Foote, C. S. (1977) in Free Radicals in Biology, II (ed. by W. A. Pryor), 85, Academic Press, New York.
4. Fridovich, I. (1976) in Free Radicals in Biology, I (ed. by W. A. Pryor), 239, Academic Press, New York.
5. Kellogg, E. W., III; Fridovich, I., J. Biol. Chem., 1975, 244, 6049.
6. Kellogg, E. W., III; Fridovich, I., J. Biol. Chem., 1977, 252, 6721.
7. Rosen, H.; Klebanoff, S. J., J. Exp. Med., 1979, 149, 27.
8. Packer, J.; Willson, R., Nature, 1979, 278, 737.
9. Anwar, M.; Prebble, J., Photochem. Photobiol., 1977, 26, 475.
10. Epel, B. L., Photophysiol., 1973, 8, 209.
11. Stoien, J. D.; Wang, R. J., Proc. Natl. Acad. Sci. USA, 1974, 71, 3961.
12. Sulkowski, E.; Genria, B.; Defaye, J., Nature, 1964, 202, 39.
13. Litwin, J., J. Gerontol., 1972, 7, 381.
14. Pereira, O. M.; Smith, J. R.; Packer, L., Photochem. Photobiol., 1976, 24, 237.
15. Packer, L.; Fuehr, K., Nature, 1977, 267, 423.
16. Packer, L.; Fuehr, K.; Walton, J.; Aggarwal, B.; Avi-Dor, Y., Energy and Environment Annual Report, Lawrence Berkeley Laboratory, 1975, 92.
17. Cheng, L. Y. L.; Packer, L., FEBS Letters, 1978, 97, 124.

18. Aggarwal, B. B.; Quintanilha, A. T.; Cammack, R.; Packer, L., Biochim. Biophys. Acta, 1978, 502, 367.
19. Cheng, L. Y. L.; Kellogg, E. W., III; Packer, L., manuscript in preparation.
20. Chance, B.; Hess, B., Science, 1959, 129, 707.
21. Porshad, R.; Sanford, K. K.; Jones, G. M.; Tarone, R. E., Proc. Natl. Acad. Sci. USA, 1978, 75, 1830.
22. Krinsky, N. I., Photochem. Photobiol., 1974, 20, 65.
23. Wu, G.-S.; Stein, R. A.; Mead, J. F., Lipids, 1978, 13, 517.
24. Maguire, J.; Packer, L., unpublished observation.

RECEIVED October 18, 1979.

Chemical Deterioration of Muscle Proteins During Frozen Storage

JUICHIRO J. MATSUMOTO

Department of Chemistry, Sophia University, Kioi-cho 7,
Chiyoda-ku, Tokyo, Japan 102

Frozen storage of meat, poultry meat and fish is one of the most important preservation methods for these foods. During frozen storage, deteriorations due to putrefaction and autolysis are decreased and the foods are satisfactorily preserved from the hygienic point of view. However, several undesirable changes still occur in the frozen stored meats.

Changes in quality of the frozen meats are demonstrated by several features. In frozen and thawed raw meat, there is increased loss of water, some changes in flavor and taste and an undesirable softening. When the frozen and thawed meat is cooked, the succulence and water-holding capacity is decreased and there are undesirable changes in texture such as toughness, coarseness and dryness. As compared with unfrozen fresh meat, the functional properties such as emulsifying capacity, lipid-binding properties, water-holding or hydrating capacity and gel capacity are decreased in the frozen stored meat.

In order to ovecome these defects, much research has been done in an attempt to clarify the mechanisms and causes of these changes during frozen storage of meats. Such studies have included a wide variety of animals including beef animals, hogs, poultry, fish, shellfish and other invertebrate aquatic animals. The number of papers published in this area amounts to several hundred.

Most of the studies indicate that denaturation of muscle proteins plays the dominant role in the quality changes of the frozen stored meats. The muscle proteins of fish and other aquatic animals have been found to be much less stable than those of beef animals, pigs and poultry (1). The present paper will be limited primarily to fish muscle as one representative of vertebrate muscle and it will also deal primarily with the behavior of fish proteins at sub-zero temperatures. In order to do a thorough analysis within the space limit permitted, focus will be on the changes of the proteins per se leaving peripheral problems to other reviews (2-18).

0-8412-0543-4/80/47-123-095$07.25/0

Structure and Composition of Muscle

Anatomical structure. The striated muscles of vertebrates consist of bundles of muscle fiber cells covered with connective
tissue. A muscle fiber is composed of bundles of striated myofibrils surrounded by sarcoplasmic reticulum, mitochondria and
other organelles. The unique structure of the striated myofibrils, composed of thin(actin)-filaments and thick(myosin)-
filaments, is common to all striated muscles of vertebrates including fish (19,20).

Chemical constitution. The approximate composition of mammalian
muscle is: 16-22% protein; 1.5-13% lipid; 0.5-13% carbohydrate;
∿1% inorganic matter; and 65-80% water (21). Poultry muscle
contains less lipid. The composition of fish muscles is: 15-24%
protein; 0.1-22% lipid; 1-3% carbohydrate; 0.8-2% inorganic
matter; and 66-84% water (21-23).
      Vertebrate muscles contain similar types of proteins, al-
though some differences do exist in their relative amounts and in
the properties of each protein. The proteins are classified into
three groups based on solubility (22,24). These are: 1) the
sarcoplasmic proteins, extractable at low ionic strength (<0.15),
which include many soluble proteins (mainly enzymes) of the
sarcoplasmic fluid, other globular proteins of organelles and
proteins attached on the sarcolemma; 2) the myofibrillar proteins, extracted by salt solutions of high ionic strength (>0.5),
which include actin, myosin, tropomyosin and troponin. In muscle, F-actin filaments, a polymerized form of monomer G-actin,
and tropomyosin and troponin, the regulatory proteins, compose
the thin filaments, while orderly aggregated myosin molecules
form the thick filaments; 3) the stroma proteins - the residual
group which is not extracted by salt solutions or dilute alkaline
or acid solutions. This group includes collagen and elastin, the
connective tissue proteins, and a new protein, connectin, found
in the fine structure of the thin filaments (25). The approximate protein compositions of various animal muscles are shown in
Table I (26-32).
      The amount of stroma proteins is less in fish muscles (3-5%)
than it is in beef or rabbit muscles (15-18%). This may explain
why raw fish fillets are acceptable in Japanese dishes, whereas
beef, rabbit and pork are rarely served raw. According to
Fennema et al. (9), tenderness is primarily related to collagen
content, while toughness and water-holding capacity are associated with the myofibrillar proteins. Many papers on cooked meat
mention both tenderness and toughness, while those on cooked fish
note the problems of toughness rather than tenderness. This also
might be related to the difference in content of the stroma
proteins.

Table I. Protein Composition of Muscle

	Sarcoplasmic proteins	Myofibrillar proteins	Stroma proteins	Reference
	(% of total proteins)			
Beef	∿17	∿68	15	(26)
Rabbit	∿27	∿57	∿15-18	(26,27)
Chicken (young)	33	62	5	(28)
Fish Teleosts	∿17-25	∿70-80	∿3-5	(29,30)
Elasmo-branchs	∿17-21	∿71-73	∿9-10	(31,32)

Properties of muscle proteins.

Myosin. Rabbit muscle myosin is a long, thin molecule (∿1400 X 20-50 Å) with a molecular weight of ∿5 X 10^5. It is composed of two heavy chains and four light chains as demonstrated by SDS-polyacrylamide disc gel electrophoresis. On tryptic digestion, myosin is split into the subunits, H-meromyosin (HMM) and L-meromyosin (LMM). HMM is further split into S-1 and S-2 subunits. While LMM is a rod of ∿90% α-helical content, the α-helical content for HMM, S-1 and S-2 fragments is 46%, 33% and 87%, respectively. The ATPase activity is localized in the S-1 subunit (33,34). Although fish myosins appear to have the same structural profile (10,22,35-40) and similar amino acid composition as rabbit myosin (39,41,42), fish myosin is different from rabbit myosin in physicochemical properties such as solubility, viscosity and stability (10,22,35-40).

Actin. Rabbit muscle G-actin is globular with a molecular weight of 4.2 X 10^4. In the presence of salts it is polymerized into F-action (34). The principal properties of fish actin (35-37,40, 43,44), including amino acid composition (41), are similar to rabbit actin, but fish actin is more readily extracted from wet muscle by salt solutions as a viscous solution of actomyosin (22,35,36,45).

Actomyosin. At high salt concentrations (e.g. 0.6 M KCl), actin
and myosin combine to form actomyosin filaments giving a highly
viscous solution. Actomyosin retains the ATPase activity of myosin and demonstrates "super-precipitation" on the addition of ATP
(24,34). As expected, there are differences between actomyosins
of rabbit and fish with respect to solubility (10,22,35,36),
viscosity (46) and ultracentrifugal behavior (47). Since actomyosin is the most readily available form of myofibrillar proteins from fish muscle, its behavior relative to deterioration
during frozen storage has been most frequently studied.

Comparative stability. Connell has demonstrated that cod myosin
(35-38,40) is much more labile and is aggregated much faster than
rabbit myosin (48,49) during storage in solution at 0°C. Buttkus
reported similar results for trout myosin. Aggregation was
attributed to formation of disulfide bond crosslinkages (50,51).
Connell (52) has also shown that myosins of several fish species
are less resistant to treatment with urea or guanidine·HCl than
myosins of rabbit, beef and chicken as determined by ultracentrifugal analyses, specific rotation, free SH groups, and ATPase
activity (Figure 1). Fish myosins were more readily digested by
trypsin than myosins of warm blooded animals (38,52). The viscosity profiles of fish actomyosin were more variable than rabbit
actomyosin during frozen storage (46,53). Fish muscle actomyosin
ATPase activity was lost more rapidly on heat treatment than
those of the mammals (54,55).
       These data indicate that fish muscle proteins are much more
fragile in structure and are much more susceptible to denaturing
factors such as storage at low temperature, heating and enzymic
treatments and exposure to chemical denaturants than mammalian
muscle proteins. Therefore, careful study of fish muscle proteins should be useful in understanding the "how and why" of
changes which occur in proteins during frozen storage.

Freeze denaturation of muscle proteins

Early investigations. After Finn (56) demonstrated that insolubilization of press juice proteins (sarcoplasmic proteins) occurred during frozen storage, Reay and Kuchel (57) reported that,
during frozen storage of haddock, the salt-soluble proteins became insoluble while the water-soluble proteins remained soluble.
This was confirmed by more comprehensive studies by Dyer and his
colleagues (58-60) on cod and other fish, showing that the amount
of extracted (native) actomyosin decreased with increased length
of storage, while the amount of the non-actomyosin (sarcoplasmic)
proteins did not show any significant change throughout frozen
storage. Since the decrease in soluble actomyosin correlated
well with palatability scores, it was proposed that denaturation
of actomyosin is the major cause for the decrease in eating
quality of stored frozen fish. Following Dyer's work, most

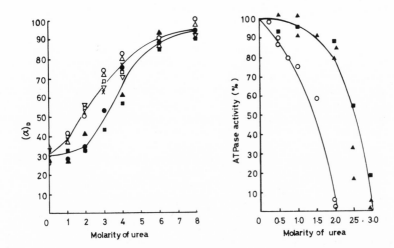

The Biochemical Journal

Figure 1.   Denaturation of myosins from various animals on treatment with urea solutions. [α]$_D$, *specific rotation;* ●, *rabbit;* ■, *ox;* ▲, *chicken;* ○, *cod;* □, *haddock;* △, *lemon sole;* ×, *plaice;* ▽ *halibut (52).*

workers have focused their attention on the myofibrillar proteins (10-17,61-63).

Information on the molecular changes occurring during frozen storage of whole muscle or of isolated protein preparations will be reviewed here.

Actomyosin. Frequently, the change in amount of soluble actomyosin is regarded as the primary criterion of freeze denaturation. It must be remembered that solubility data do not tell precisely how much protein is denatured and how much is native; rather, it provides a relative measure of denaturation. Solubility decreases have been found in frozen storage experiments with either intact muscle, protein solutions or with suspensions of isolated actomyosin.

Viscosity of soluble actomyosin fractions decreased with increasing time of storage (64-68). This suggests that the actomyosin filaments have become less rod-like or less filamentous either by individual molecular folding or by aggregation of the filaments.

As noted by ultracentrifugal analyses of the soluble fractions, the actomyosin peak (20S-30S) decreased in area, while several faster moving peaks simultaneously appeared with increasing time of frozen storage (47,66,68). It has been proposed that actomyosin forms various aggregated states during frozen storage (11,15,17,50,51,61,66-68), which is in agreement with the viscosity change.

On the other hand, King (69) and Anderson and coworkers (70,71), based on detailed analyses of ultracentrifugal patterns of extracts of frozen stored cod muscle and experiments on the effect of lipids on protein denaturation, have proposed that denaturation of F-actomyosin occurs by two parallel pathways which lead to insolubilization (Figure 2). As indicated by Connell (61), the occurrence of G-actomyosin at an intermediary stage needs experimental verification. Possibility of an alternate pathway involving lipids will be discussed later.

Electron microscopic analyses of isolated preparations of fish actomyosin denatured by frozen storage (68,72-74) showed that actomyosin filaments with arrowhead structures aggregated side-to-side and crosswise when thawed immediately after freezing. As time of frozen storage increased further aggregations formed network structures (Figure 3).

In addition to aggregation, dissociation of F-actomyosin into F-actin and myosin also occurred. It appeared that the dissociated F-actin, as thin filaments, became entangled and aggregated and that the dissociated myosin monomers folded into globular form. At advanced stages of freeze denaturation, large masses with diffuse outlines were frequently found suggesting complex aggregation of actin and myosin.

The above proposal is summarized in Figure 2C. This model appears to be consistent with solubility, viscosity, and ultra-

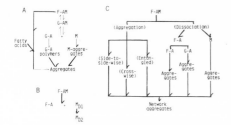

Figure 2. Hypothetical mechanisms of aggregation of fish actomyosin during frozen storage. (A) King, 69; (B) Connell, 61; (C) Matsumoto (* proposal in the present paper). AM, actomyosin; M, myosin; M_{D1} and M_{D2}, denatured myosin; A, actin.

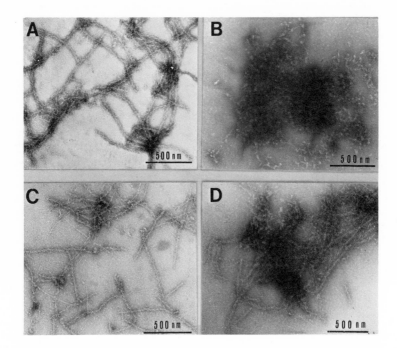

Journal of Biochemistry

Figure 3. Electron micrographs of carp actomyosin before and after frozen storage in 0.05M KCl at $-20°C$. A and B, no additives; C, 0.2M sodium glutamate added; D, 1M glucose added. A, before freezing; B, C and D, after 8, 9 and 9 weeks of frozen storage, respectively. Each specimen was negatively stained with uranyl actate solution (72).

centrifugal data which indicate that a decrease in asymmetry and progressive aggregation occur during frozen storage.

Jarenbäck and Liljemark (75,76) found similar changes in cod actomyosin solution and cod muscle during frozen storage. The denatured myosin was not extracted with salt solution.

The above mentioned dissociation of actomyosin into actin and myosin could be due to a shift in the equilibrium, actomyosin ⇌ actin + myosin, by the highly concentrated salt solution of the unfrozen liquid portion in the protein-water system (22, 77). However, if this is true, the dissociated actin and myosin must re-associate immediately after thawing. This may be difficult since the ability to associate is decreased during frozen storage.

ATPase activity, another property of myosin related to its contractile function, as is the actin-binding property, is also decreased by frozen storage. The specific ATPase activity of fish actomyosin decreases with increased time of frozen storage (66,67,72,78-82). This decrease should be due to a decrease in the ATPase activity of myosin. The rate of decrease is slower than that of free myosin (80,82). Connell (78) and Kawashima et al. (83) have detected some ATPase activity in insoluble aggregated actomyosin.

During frozen storage of muscle, isolated actomyosin and myosin changes have been sought in the number of -SH groups (50, 51,78,84), titratable acid groups (85), and net charge (86), and in the salting-out profiles (46,47,53,65,68,87,88). Significant changes have been found only in the salting-out profiles.

When 0.1 M sodium glutamate was added to carp actomyosin, denaturation during frozen storage was almost eliminated, as measured by changes in solubility, viscosity, ultracentrifugal behavior, ATPase activity and electron microscopic profiles (66,72) (Figure 3). This protective effect of sodium glutamate will be discussed below.

Connell has proposed that insolubilization of actomyosin during frozen storage of cod muscle is attributable to the denaturation of myosin rather than actin (89). During 40 weeks storage at -14°C, extractability of actomyosin and myosin decreased in parallel, while that of actin appeared to remain constant. The decrease in extractability of myosin was biphasic, while that of actomyosin followed an exponential curve.

However, our work on in vitro frozen storage of isolated carp actomyosin showed that actin is denatured progressively with myosin as demonstrated by SDS-polyacrylamide disc gel electrophoresis (90).

Analysis of the rates of insolubilization of isolated carp actomyosin indicated that denaturation proceeds by two or more first order processes with different rate constants (73). During the initial rapid stage, it appears that both myosin and actin undergo denaturation, while during the second stage, tropomyosin and troponin undergo denaturation (90).

Myosin. Because of the difficulty of isolating pure myosin from
fish, studies on the behavior of this protein during frozen storage were delayed. Connell (91) was the first to conduct such
studies. Solutions of cod myosin in 0.6 M KCl were frozen and
stored at different temperatures ranging from -7 to -78°C and the
thawed solutions were examined by ultracentrifugal analysis. The
progressive aggregation of myosin monomers to dimers, trimers and
other larger polymers was demonstrated. Neither the specific rotation nor the number of active SH groups changed appreciably
during the polymerization. Aggregation was ascribed to bonding
of an unknown nature rather than to disulfide bonding. Connell
suggested that myosin molecules aggregated side-to-side without
unfolding or undergoing any change in intramolecular conformation. He also found that the amount of aggregated myosin in myosin preparations from frozen, stored cod increased with increasing time of storage (89).

Changes in the solubility, ultracentrifugal behavior, number
of SH groups and electron microscopic profiles in non-freeze
stored or frozen myosins of rabbit and trout, as observed by
Buttkus (50,51), supported those of Connell (91). The rate of
aggregation was the highest around the eutectic point (-11°C) of
the myosin-KCl-water system. Side-to-side aggregated dimers of
rabbit myosin were observed by electron microscopy, as illustrated previously by Slayter and Lowey (92).

Changes in solubility, viscosity, ATPase activity, and ultracentrifugal and salting-out profiles were found during frozen
storage at -20°C of carp myosin solutions (in 0.6 M KCl) and carp
myosin suspensions (in 0.05 M KCl) (82,93).

Myosins isolated from various frozen stored fish muscles had
slightly lower ATPase activity than those from fresh muscles (94,
95). A decrease in ATPase activity was found also with isolated
carp myosin when stored at -20°C (82); the rate of decrease was
faster than with actomyosin isolated from the same fish. As with
actomyosin, the decrease in ATPase activity was preceded by a
temporary rise in activity (~150% the pre-freezing value).

Like rabbit myosin (96), isolated carp myosin forms filaments at low ionic strength (0.05) which are observable with the
electron microscope at moderate magnification. These filaments
are either "spindle shaped" (dialysis against low ionic strength
buffer) or "dumbbell shaped" (immediate dilution to low ionic
strength).

After frozen storage of myosin solutions and suspensions of
the filaments, reconstitution of filaments was attempted. The
filament suspensions were thawed, dissolved in 0.6 M KCl, and
then examined for reconstitution at low ionic strength. Filaments formed from either frozen-stored samples (filaments or myosin solution) were not as perfect in shape as those prepared from
unfrozen, intact myosin. The spindle-shaped myosin was more stable in frozen storage than the dumbbell-shaped myosin. Dissolved

myosin was least able to form filaments ($\underline{82}$) after frozen storage (Figure 4).

When 0.1 $\underline{M}$ sodium glutamate was added to solutions of myosin prior to freezing, solubility, viscosity, ATPase activity and filament-forming capacity remained at the level observed before frozen storage ($\underline{82}$).

Subunits of myosin. Hanafusa ($\underline{97},\underline{98}$) found that isolated rabbit myosin and HMM underwent denaturation when cooled to temperatures ranging from -10° to -196°C followed by immediate thawing. Denaturation was measured by an increase in viscosity, change in absorbance at 278 nm and change in the optical rotatory dispersion coefficients, a_0 and b_0, in the Moffit-Young equation. The absolute value of b_0 decreased while that of a_0 increased with a decrease in the freezing temperature.

In the author's laboratory, HMM and LMM were obtained by trypsin digestion. LMM was purified by reprecipitation in 75% ethanol and redissolved in 0.5 $\underline{M}$ KCl-0.05 $\underline{M}$ tris-maleate buffer, pH 6.5, followed by ultracentrifugation ($\underline{39}$).

HMM and LMM were stored frozen at -20°C, and the changes in properties were followed for each protein ($\underline{82}$). While there were no significant changes in the solubility curves for HMM and LMM, appreciable changes were found in other properties. ATPase activity of HMM decreased to 50% of the pre-freezing value after 1 day and was not detectable after 2 weeks. No initial increase in activity was found with HMM. The rate of decrease in ATPase activity was much faster than with myosin solutions, where about 55% of the initial activity was retained after 7 days frozen storage. The ability of HMM to bind with F-actin, as determined by electron microscopy, was lost after 2 weeks frozen storage.

LMM, after dialysis against a solution of 0.05 $\underline{M}$ KCl-0.005 $\underline{M}$ tris-maleate buffer (pH 6.2), also exhibited a decreased capacity to form well-ordered paracrystals or tactoids as examined by electron microscopy (Figure 5). It is interesting that several half-reconstituted paracrystals (Figure 5b) were found after 6 weeks of frozen storage.

These results indicate that denaturation of myosin probably occurs in both the HMM and LMM segments. The decrease of ATPase activity and F-actin binding capacity of HMM may indicate a conformational change in the light chains which are located in the S-1 region of HMM. Based on a lack of an observed change in specific rotation, Connell ($\underline{91}$) has argued against the possibility of a conformational change of myosin molecules during frozen storage. However, a conformational change in the light chains of HMM may not necessarily be accompanied by an appreciable change in specific rotation. A change from an ordered random coil to a disordered random coil appears to have occurred during frozen storage of HMM and myosin. If so, an appreciable change in specific rotation would not be expected.

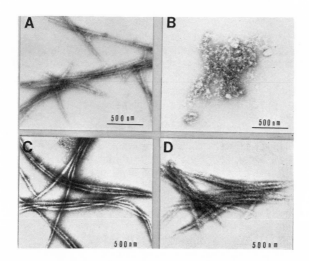

Figure 4. Electron micrographs of reconstituted spindle-shaped filaments of carp myosin before and after frozen storage in 0.05M KCl at −20°C. A and B, no additives; C and D, 0.2M sodium glutamate added. A and C, reconstituted before freezing; B and D, reconstituted after 2 and 6 weeks of frozen storage, respectively stained (82).

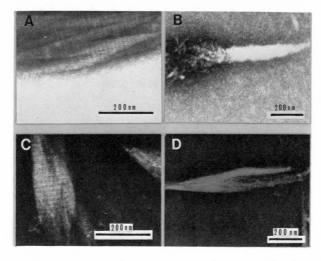

Figure 5. Electron micrographs of reconstituted paracrystals of carp LMM before and after frozen storage in 0.6M KCl at −20°C. A and B, no additives; C and D, 0.2M sodium glutamate added. A and C, reconstituted before freezing; B and D, reconstituted after 8 weeks of frozen storage. Negatively stained. Frozen storage in 0.05M KCl gave similar results (82).

The decrease of paracrystal-forming capacity of LMM during
frozen storage might be explained by either an impairment of the
highly ordered helical structure needed for the orderly alignment
in a paracrystal, or by disorderly side-to-side aggregation of
two or more myosin molecules preventing the orderly alignments in
a paracrystal. Practically, both are likely to happen during denaturation in the frozen state.

Addition of 0.1 M sodium glutamate reduced the changes occurring in HMM and LMM during frozen storage (Figure 5).

Actin. Connell (89) did not find a significant amount of denaturation of actin in frozen cod muscle during storage for 100
weeks. In contrast, the author and his colleagues found that
denaturation occurred during frozen storage of isolated carp
actin (82,99). Actin was prepared by the method of Guba-Straub
(24) modified by use of buffer A of Spuchdich-Watt (100,101) for
the extraction. Actin in either G- or F-form was frozen and
stored at -20°C. Its solubility, viscosity, polymerizing ability
(G-actin), and profiles by electron microscopy (F-actin) were
tested at intervals during frozen storage. Both G- and F-actins
showed denaturation during frozen storage. Solubilities of both
G- and F-actins and viscosity of F-actin decreased with storage
time. Polymerizing ability of G-actin, as indicated by an increase in viscosity after addition of $MgCl_2$, NaCl or KCl, decreased gradually. While freshly prepared F-actin showed fine
thin filament structures by election microscopy, F-actin stored
frozen in 0.05 M NaCl for 4 weeks at -20°C showed aggregates of
entangled filaments with vague outlines (Figure 6). Such de-
formed filaments were similar to those found in the denaturation
of actomyosin during frozen storage (Figure 3).

Loss of the polymerizing ability of G-actin molecules indicates that the conformation of native G-actin, a globular molecule, must have been impaired during frozen storage.

Freezing and storage after addition of sodium glutamate decreased the rate of denaturation. The solubility did not decrease and the F-actin filaments kept their fine structures during frozen storage (Figure 6).

Tropomyosin and troponin. Tropomyosin is apparently the most
stable of the fish fibrillar proteins during frozen storage. It
can be extracted long after actin and myosin become inextract-
able; however, it does denature gradually (90).

Troponins isolated from frozen-stored bigeye tuna, Tilapia
or Beryx, were less active in their regulatory function than
those from fresh muscle (102).

Myofibrils and tissues. Myofibrils, a systematically organized
complex of myofibrillar proteins, undergo some structural changes
during frozen storage of fish muscle. The most noticeable
changes during storage are the fusion of the myofibrils as illus-

trated by the cell fragility method (103,104) and fragmentation
into short pieces at the Z-bands (76,105,106). These changes
have been discussed elsewhere (16).

Connective tissue proteins. Collagen comprises the major material of skin, myocommata and sarcolemma. In a study of "gaping",
in which slits and holes appear and sometimes the fillet falls
apart, Love and his co-workers have attributed this defect to the
behavior of the myocommata proteins (107,108). This problem has
been reviewed elsewhere (16).

Sarcoplasmic proteins and other proteins. Since Reay and Kuchel
(57) and Dyer (58) discovered that denaturation of myofibrillar
proteins is of such profound importance in the toughness of fish,
little attention has been given to the water-soluble sarcoplasmic
proteins, which include enzymes and other proteins in the sarco-
plasma fluid, subcellular organelles and cell membranes. More
recently, papers have appeared on the denaturation of enzymes
during frozen storage (81,109-121). These studies have demonstrated that catalase, alcohol dehydrogenase, glucose dehydrogenase, lactate dehydrogenase and malate dehydrogenase from various
sources other than fish lose enzymic activity during frozen storage of the solutions. In a study by the author and his colleagues where the enzyme solutions were stored at -20°C (82,122),
inactivation was more marked at lower enzyme concentrations,
while at high concentrations inactivation was essentially zero
(Figure 7). No insolubilization of the enzymes was noted. As
discussed by Hanafusa (97,98,113-116), inactivation of enzymes
which are globular might involve unfolding of the intramolecular
structure.

Inactivation of enzymes during frozen storage was prevented
by addition of 0.2 $\underline{M}$ sodium glutamate, 0.2 $\underline{M}$ $(NH_4)_2SO_4$, or 0.1%
egg albumin; a synergistic cryoprotective effect was illustrated
between sodium glutamate and albumin (122).

Some enzymes and enzyme systems are still active at the temperature of frozen storage (123-132). Such enzymatic activity,
especially of proteases, may cause loss of biological activity of
actomyosin and other muscle proteins. Products of such enzymatic
activity, e.g. free fatty acids and formaldehyde, may effect a
secondary denaturation of muscle proteins.

Crosslinkages. Connell argued against a role for disulfide bond
formation in the intermolecular aggregation of myosin (91) since
he did not detect any significant change in the number of free SH
groups of the whole macerate of frozen stored cod (78,91). Addition of 1% sodium dodecylsulfate (SDS) solubilized almost all the
myofibrillar proteins of cod flesh stored for up to 29 weeks at
-14°C. Therefore, Connell attributed the crosslinkages in the
severely toughened cod flesh to formation of non-covalent bonds
(133). In contrast, involvement of SH groups in the process of

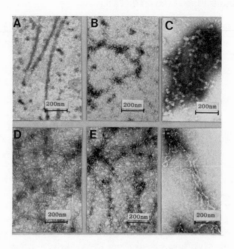

Figure 6. *Electron micrographs of F-actin filaments of carp before and after
frozen storage in 0.05M KCl at −20°C. A, B and C, no additives; D, E and F,
0.2M sodium glutamate added. A and D, before freezing; B and E, after 1 week
of frozen storage; C and F, after 4 weeks of frozen storage. Negatively stained*
(82).

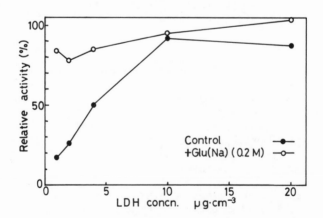

Figure 7. *Decrease of LDH activity during frozen storage for 16 hours at
−20°C, in the absence and presence of 0.2M sodium glutamate. −●−, no additives; −○−, 0.2M sodium glutamate added* (122).

denaturation of trout and rabbit myosins during frozen storage
has been emphasized by Buttkus (50,51). He has successfully re-
solubilized aggregated myosin by using a solution containing both
6 M guanidine·HCl and either 0.5 M mercaptoethanol, 0.3 M sodium
sulfite or 0.3 M sodium cyanide. From these results, he has attributed the crosslinkages involved in aggregation of myosin during frozen storage to disulfide bonds, hydrophobic bonds and hydrogen bonds. Free SH groups might be oxidized first to disulfide bonds. However, only a small decrease was found in the number of free SH groups during frozen storage. Therefore the
changes appear to be the result of rearrangements of disulfide
bonds from intramolecular to intermolecular ones through a sulf-
hydryl/disulfide interchange reaction. Although it is based on
work on plant cells, Levitt's theory on the important role of disulfide bond formation during freeze injury should be noted
(134).

Experiments on crosslinking have been carried out in the
author's laboratory (93). Solutions of isolated carp actomyosin
or myosin in 0.6 M KCl or suspensions in 0.05 M KCl have been
stored at -20°C. Samples were taken at intervals and homogenized
with several different solutions. The solutions used were water
(to test for nonspecific association forces), 0.6 M KCl (to test
for ionic bonds), 0.5 M β-mercaptoethanol (to test for disulfide
bonds), 1.5 M urea (to test for hydrogen bonds), 8 M urea (to
test for hydrogen bonds and nonpolar bonds), and 1 M KOH (to test
for ionic bonds and others) (135,136). Combinations of these
solvents were also tested.

The results with carp actomyosin after various times of frozen storage are shown in Figure 8. Different solvents gave different results. Myosin gave similar patterns.

Much poorer solubilities were obtained with solvents which
did not contain KCl. Complete resolubilization was not accomplished with only 0.5 M β-mercaptoethanol and 8 M urea.

These results led to the conclusion that denaturation and/or
insolubilization of actomyosin and myosin during frozen storage
is a result of aggregation caused by the progressive increase in
intermolecular crosslinkages due to formation of hydrogen bonds,
ionic bonds, hydrophobic bonds and disulfide bonds.

Effect of cryoprotective agents

Cryoprotective agents which lessen denaturation of proteins
during frozen storage have been found to include sugars, polyalcohols, and compounds of other families. These protective effects have been studied not only in the preservation of foods but
in storing of microorganisms and of biological materials such as
enzymes, vaccines, blood and organs (110-116,137).

The first published use of cryoprotectants for muscle proteins, found to be successful in commercial application, was a
combination of sucrose (10%) and polyphosphate (0.2-0.5%) which

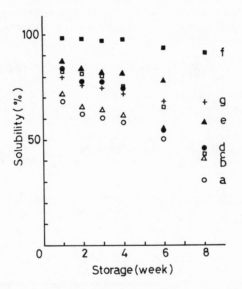

Figure 8. Changes in solubility of carp actomyosin in various solvents following frozen storage at −20°C (93). Composition of the solvents: a (○), 0.6M KCl; b (△), 0.6M KCl + 1.5M urea; c (□), 0.6M KCl + 8M urea; d (●), 0.6M KCl + 0.5M β-mercaptoethanol (ME); e (▲), 0.6M KCl + 1.5M urea + 0.6M ME; f (■), 0.6M KCl + 8M urea + 0.5M ME; g (+), 0.6M KCl + 0.2M KOH.

prevented denaturation of muscle proteins of Alaska pollack, a fish very sensitive to frozen storage (138-140). Thorough washing of the minced muscle with water prior to addition of the cryoprotectants was necessary to prevent denaturation during frozen storage (141). Half of the sucrose may be replaced by sorbitol in the cryoprotectant mixture.

Cryoprotectants which have so far been found to be effective for fish muscle proteins include such compounds as monosaccharides, oligosaccharides, polysaccharides of relatively small molecular size, di- and polyalcohols, hydroxymonocarboxylic acids, di- and tricarboxylic acids, acidic, basic and some other amino acids, and phosphates and their derivatives (15,16,66,67,72-74, 82,83,93,97-99,112-114,122,140-154). Dimethylsulfoxide (DMSO), which is cryoprotective for various biological materials such as red cells, was not effective for fish muscle proteins (147).

Based on a systematic study of the relationships between the molecular structures of the compounds and their cryoprotective effects, the following requirements for exhibiting cryoprotective effects for fish muscle proteins have been proposed. 1) A molecule has to possess one "essential group", either $-COOH$, $-OH$, or $-OPO_3H_2$, and more than one "supplementary group", of the type $-COOH$, $-OH$, $-NH_2$, $-SH$, $-SO_3H$, and/or $-OPO_3H_2$. 2) The functional groups must be suitably spaced and properly oriented with respect to each other. 3) The molecule must be comparatively small (67, 150). Among the cryoprotective agents studied by the author and his colleagues, sodium glutamate was most effective. Addition of 0.1-0.2 $\underline{M}$ sodium glutamate to protein solutions or suspensions prior to freezing and storage prevented denaturation of carp actomyosin as measured by the methods described earlier. Effectiveness of sodium glutamate in improving the quality of meat and frozen meat has been reported by Norton et al. (151).

The cryoprotective effects of sodium glutamate have been shown for carp myosin, its subunits, and actin as presented above. It should be noted that sodium glutamate exists primarily as $^\ominus OOC-CH_2-CH_2-CH(NH_3^\oplus)-COO^\ominus$ at $\sim$ pH 7 where the frozen storage experiments on the isolated proteins were done.

Mechanism of freeze denaturation and cryoprotective effect

Cause of denaturation. Many hypotheses have been proposed to explain the denaturation of muscle proteins (9-17). These hypotheses include: 1) the effects of inorganic salts concentrated into the liquid phase of the frozen system; 2) water-activity relations; 3) reactions with lipids; 4) reaction with formaldehyde derived from trimethylamine (in fish); 5) auto-oxidation; 6) surface effects at the solid-gas interface; 7) effects of heavy metals; and 8) effects of other water-soluble proteins (such as proteases).

Among the above hypotheses, effects of lipids (4-17,59-62, 69-71,155-159), formaldehyde (160-166), and gas-solid interface (167) appear to be very important in Gadoid fishes. Denaturation of myofibrillar proteins caused by free fatty acids and/or lipid peroxides must occur during frozen storage. To prove this, Jarenbäck and Liljemark have shown by electron microscopy that, in muscle stored frozen with added linoleic and linolenic hydroperoxides, myosin became resistant to extraction with salt solution (168).

However, recent results on isolated protein preparations show that proteins undergo denaturation in the absence of lipids, formaldehyde, heavy metals and water-soluble proteins.

Another popular view is the so-called "salt-buffer hypothesis" which gives attention to the effects of highly concentrated salt solution in the unfrozen phase on frozen muscle proteins. The concentrated salt solution may denature the proteins (9-17, 169-177). Whereas experiments with isolated muscle protein preparations cannot exclude the effects of salts such as NaCl or KCl (since they are required to solubilize the proteins), denaturation during frozen storage has been decreased or prevented completely when an efficient cryoprotectant such as sodium glutamate or glucose was added (66,67,82,93,145-150). Hence, the effect of salts may not be of primary importance, though they may contribute.

Other factors, except for water-activity relations, might take place in some fish species or in some conditions of freezing and storage. Nevertheless, these do not appear to be of general importance because denaturation occurs in experiments on protein solutions where such factors are excluded.

The water-activity relations, effects of displacements of water or effects of changes in the state of water must be the most important factors to trigger and to promote the denaturation of muscle proteins during frozen storage.

As described by Fennema (9), several refined hypotheses such as "physical barrier and structured water hypothesis" (134,178, 179), "ice-moderator hypothesis" (180-183), and "minimum cell volume hypothesis" (184) have been proposed. However, the author will take a more naive approach in interpreting the results on denaturation of muscle proteins during frozen storage at the same time taking advantage of the basic ideas of the above hypotheses.

Aggregation of myosin and LMM.

Data on myosin (50,51,82,91) and LMM (82) support side-to-side aggregation of molecules without appreciable change in conformation during frozen storage, as proposed by Connell (91). A working model of the author and his colleagues (67,82,150) will be described here (Figure 9). In aggregations of this type, the myosin molecules are associated or cross-linked with each other through the tail (LMM and S-2 subunit) parts of each molecule.

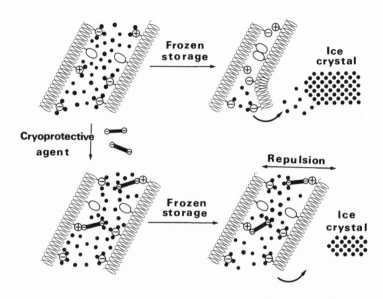

Figure 9.   A schematic model of denaturation of α-helical proteins during frozen storage and its prevention by cryoprotectants.   The case with dianionic cryoprotectants is illustrated.

This association may be caused by dehydration of the protein molecules as a result of displacement of water molecules from the hydrating sites and the surrounding area of the protein molecules. Then, the protein molecules come closer to each other resulting in a higher probability of forming intermolecular cross-linkages at any site available. The a-helical structures undergo little or no unfolding. The water molecules move in the direction of decreasing vapor pressures, giving rise to ice crystal formation. When the system is thawed, the water molecules turn to liquid water and return to the vicinity of the protein aggregates. Rehydration of the protein molecules is incomplete however, because the protein-water affinity is equal to or less than the protein-protein affinity.

If there is a cryoprotectant (with two or more functional groups) in the system prior to freezing, the cryoprotectant molecules may bind or associate with protein molecules at one of the functional groups either by ionic bonds or by hydrogen bonds. Thus each protein molecule can be coated with cryoprotectant. The cryoprotectant molecule can be hydrated at the remaining free functional groups to form a hydrated "protein-cryoprotectant complex". When the system is frozen, some of the water molecules freeze out but some remain still attached to the cryoprotectant. The protein molecules are prevented from contacting and aggregating. In case the cryoprotectant is multifunctional (more than one positive or negative group), it will alter the positive (or negative) charge on the protein molecule thereby forming negatively (or positively) charged protein-cryoprotectant complexes which have a mutually repulsive force. These complexes may resist loses of water molecules located between the protein molecules (Figure 9).

Di-, tri- and polyhydroxy compounds, such as glycerol, sugars and sorbitol have been assumed to work by the same principle as above, but this will be discussed later.

Unfolding of globular proteins and subunits. Data on frozen storage of HMM, actin and sarcoplasmic enzymes have led us to propose that denaturation involves unfolding of the protein chain based on a decrease in enzymatic activity (myosin, HMM, and sarcoplasmic enzymes), polymerizing ability (actin) and filament forming properties (myosin) (82,99,113-116,122).

To account for such changes, the author and his colleagues have proposed that these globular proteins denature through unfolding during frozen storage by the following process (Figure 10). The native conformation of the globular molecules is maintained largely by the intramolecular nonpolar bonds which have resulted from the thermodynamic balance consequence between the two systems, namely, 1) folded protein-water and 2) unfolded protein-water. In the first system, the nonpolar groups on the polypeptide backbone are oriented inward so as to avoid contact with the water phase. In the second system, some of the nonpolar groups are projected to the interface with water forming oriented structures or clusters of molecules (9,12,185,186). In the unfrozen state, the former system is more favorable. If the system is frozen, the water molecules surrounding the protein are displaced to form ice crystals and the balance in system one may be shifted in favor of a balance between the two systems, 1) folded protein with poor hydration and 2) unfolded protein with poor hydration. Now the thermodynamic factor is in favor of the second system and the protein molecules are unfolded. The native conformation is impaired and the site of enzymatic activity is disrupted.

When cryoprotectant molecules are present, prior to freezing, some of them may be associated with or bound to the protein molecules. This results in an increased hydration of the protein molecules and an increased resistance against displacement of water even when the system is frozen. These factors may result in hindering unfolding of the protein molecules which would result in aggregation (Figure 10).

Cryoprotectants as water structure modifiers

Current theories regarding the mechanism of cryoprotectant action appear to emphasize their role as modifiers of water structure which then interfere with formation of and growth of ice crystals (9,12,185,186). The present view of the author appears, therefore, rather radical. However, as far as the proposed mechanism for action of sodium glutamate and other ionic compounds are concerned, there are several facts which indicate that sodium glutamate is bound to proteins. When mixtures of carp actomyosin and sodium glutamate were centrifuged at 50,000 rpm, a considerable amount of sodium glutamate sedimented with actomyosin (187). In a differential thermal analysis of a system with sodium glutamate added to cod mince, the amount of water frozen out was as much as in the control mince, while a system with glucose added exhibited a decrease in the freezable water (188). Sodium glutamate promotes gelation of actomyosin solutions on standing at 40°C (189). Electron microscopic pictures of carp actomyosin stored frozen with sodium glutamate exhibited straight stretched filaments, while those without glutamate showed randomly curved filaments (72). The author, therefore, does not accept the water structure modifier hypothesis proposed to explain the role of sugars and polyalcohols as being applicable to sodium glutamate. Warner has suggested (190) to the author that the role of glutamate and other amino acids might be accounted for by his hexagonal lattice theory (191) on water structure. This theory might be valid for the mono-amino monocarboxylic amino acids but binding with protein appears to occur with sodium glutamate.

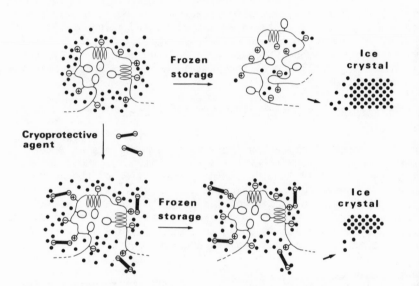

Figure 10. A schematic model of denaturation of globular proteins during frozen storage and its prevention by cryoprotectants. The case with dianionic cryoprotectants is illustrated.

Summary

Denaturation of proteins plays the major role in the deterioration of the food quality of meats of beef, pork, poultry and fish during frozen storage. Current understanding of the behavior of the muscle proteins during frozen storage has been reviewed with emphasis on the muscle proteins of fish which are the most susceptible to frozen storage among the muscle proteins of animal origin.

It has been known for a long time that denaturation of actomyosin, the main constitutent of the myofibril, occurs during frozen storage. Recent studies have shown that freezing and frozen storage denature actin and myosin, the component proteins of the actomyosin complex, and HMM and LMM, the subunits of myosin. The enzymes of the sarcoplasmic fluid also undergo denaturation during frozen storage.

Denaturation of myosin and actomyosin has so far been ascribed to intermolecular aggregation, but recent investigations have shown that intramolecular transconformation, the unfolding of the polypeptide chains, occurs in globular proteins and in subunits with globular structures.

Denaturation during storage at sub-zero temperatures may be minimized by use of suitable cryoprotective agents which include several families of compounds.

The mechanisms of denaturation during frozen storage and of the cryoprotective effects have been discussed and a hypothetical model has been presented.

Acknowledgments

The author acknowledges the help of Dr. T. Tsuchiya and colleagues in the Biochemistry Laboratory, Department of Chemistry, Sophia University, in collecting reference materials and in finishing the drawings and pictures.

Literature Cited

1. Shepherd, A. D. In "Conference on Frozen Food Quality, U.S. Dept. Agr.", 1960, ARS-74-21.
2. Bate-Smith, E. C. Advan. Food Res., 1948, 1, 1.
3. Hamm, R. Advan. Food Res., 1960, 10, 356.
4. Fennema, O.; Powrie, W. D. Advan. Food Res., 1963, 13, 220.
5. Schultz, H. W., Ed. "Symposium on Foods: Proteins and Their Reactions"; AVI: Westport, Conn., 1964, p. 472.
6. Luyet, B. J. In "The Physiology and Biochemistry of Muscle as a Food"; (Briskey, E. J.; Cassens, R. G.; Trautman, J. C., Eds.), The University of Wisconsin Press: Madison, WI, 1966, 353.
7. Hawthorn, J.; Rolfe, E. J., Eds. "Low Temperature Biology of Foodstuffs"; Pergamon Press: Oxford, 1968, p. 458.

8. Lawrie, R. A., Ed. "Proteins as Human Food"; Butterworths:
    London, 1970, p. 535.
9. Fennema, O. R.; Powrie, W. D.; Marth, E. H., Eds. "Low
    Temperature Preservation of Foods and Living Matter"; Marcel
    Dekker Inc.: New York, 1973, p. 577.
10. Dyer, W. J.; Dingle, J. R. In "Fish as Food"; (Borstrom,
    G., Ed.), Academic Press: New York, 1961, 1, 275.
11. Love, R. M. In "Cryobiology"; (Meryman, H. T., Ed.), Aca-
    demic Press: London, 1966, 317.
12. Sikorski, Z.; Olley, J.; Kostuch, S. "Critical Reviews in
    Food Science and Nutrition"; CRC Press Inc.: Cleveland,
    1976, 8, 97.
13. Partmann, W. Dechema Monographie, 1969, 63, 93.
14. Partmann, W. Z. Ernährungswiss., 1977, 16, 167.
15. Matsumoto, J. J. In "Behavior of Proteins at Low Tempera-
    tures"; (Fennema, O., Ed.), ACS Symposium Series, in press.
16. Matsumoto, J. J. In "Rheology of Foods"; Vol. 4 (Matsumoto,
    Y., Ed.), Shokuhin Shizai Kenkyu Kai: Tokyo, 1978, 83.
17. Noguchi, S. In "Proteins of Fish Muscle"; (Japanese Society
    of Scientific Fisheries, Ed.), Koseisha-Koseikaku K.K.:
    Tokyo, 1977, 99.
18. Partmann, W. In "Water Relations of Foods"; (Duckworth, R.
    B., Ed.), Academic Press: London, 1975, 505.
19. Huxley, H. E. In "The Structure and Function of Muscle";
    2nd Edition, Vol. 1, Part A (Bourne, G. H., Ed.), Academic
    Press: New York, 1972, 301.
20. Love, R. M. "The Chemical Biology of Fishes"; Academic
    Press: New York, 1970, p. 545.
21. Forrest, J. G.; Aberle, E. D.; Hedrick, H. B.; Judge, M. D.;
    Merkel, R. A. "Principles of Meat Science"; W. H. Freeman &
    Co.: San Francisco, 1975, p. 417.
22. Hamoir, G. Advan. Protein Chem., 1955, 10, 227.
23. Jaquot, R. In "Fish as Foods"; Vol. 1 (Borgstrom, G., Ed.),
    Academic Press: New York, 1961, 145.
24. Bailey, K. In "The Proteins"; Vol. 2, Part B, Academic
    Press: New York, 1954, 951.
25. Maruyama, K.; Fujii, T.; Kuroda, M.; Kikuchi, M. J.
    Biochem., 1975, 77, 767.
26. Bate-Smith, E. C. Proc. Roy. Soc. London, 1937, B124, 136.
27. Weber, H. H.; Meyer, K. Biochem. Z., 1933, 266, 138.
28. Robinson, D. S. Biochem. J., 1952, 52, 621.
29. Dyer, W. J.; French, H. V.; Snow, J. M. J. Fish. Res. Bd.
    Can., 1950, 7, 585.
30. Migita, M.; Matsumoto, J. J.; Aoe, N. Bull. Japan. Soc.
    Sci. Fish., 1959, 24, 751.
31. Shimizu, Y.; Simidu, W. Bull. Japan. Soc. Sci. Fish., 1960,
    26, 806.
32. Subba-Rao, G. N. In "The Biochemistry of Fish"; (Williams,
    R. T., Ed.), Biochemical Society Symposium No. 6, Cambridge
    University Press: London, 1950, 8.

33. Lowey, S.; Slayter, H. S.; Weeds, A. G.; Baker, H. <u>J. Mol.</u>
    <u>Biol.</u>, 1969, 42, 8.
34. Ebashi, S.; Nonomura, Y. In "The Structure and Function of
    Muscle"; 2nd Edition, Vol. 3 (Bourne, G. H., Ed.), Academic
    Press, New York, 1973, 285.
35. Connell, J. J. <u>Biochem. J.</u>, 1954, 58, 360.
36. Connell, J. J. <u>Biochem. J.</u>, 1958, 69, 5.
37. Connell, J. J. <u>Biochem. J.</u>, 1958, 70, 81.
38. Connell, J. J. <u>Biochem. J.</u>, 1960, 75, 530.
39. Tsuchiya, T.; Matsumoto, J. J. <u>Bull. Japan. Soc. Sci. Fish.</u>,
    1975, 41, 1319.
40. Connell, J. J. <u>Biochim. Biophys. Acta</u>, 1963, 74, 374.
41. Connell, J. J.; Howgate, P. F. <u>Biochem. J.</u>, 1959, 71, 83.
42. Buttkus, H. <u>J. Fish. Res. Bd. Can.</u>, 1967, 24, 1607.
43. Seki, N. In "Proteins of Fish"; (Japanese Society of Scien-
    tific Fisheries, Ed.), Koseisha-Koseikaku K.K.: Tokyo, 1977,
    7.
44. Dingle, J. R. <u>J. Fish. Res. Bd. Can.</u>, 1959, 16, 243.
45. Roth, E. <u>Biochem. Z.</u>, 1949, 318, 74.
46. Migita, M.; Suzuki, T. <u>Bull. Japan. Soc. Sci. Fish.</u>, 1959,
    25, 127.
47. Suzuki, T.; Kanna, K.; Tanaka, T. <u>Bull. Japan. Soc. Sci.</u>
    <u>Fish.</u>, 1964, 30, 1022.
48. Holtzer, A. ·<u>Arch. Biochem. Biophys.</u>, 1956, 64, 507.
49. Holtzer, A.; Lowey, S. <u>J. Am. Chem. Soc.</u>, 1956, 78, 5954.
50. Buttkus, H. <u>J. Food Sci.</u>, 1970, 35, 558.
51. Buttkus, H. <u>Can. J. Biochem. Physiol.</u>, 1971, 49, 97.
52. Connell, J. J. <u>Biochem. J.</u>, 1961, 80, 503.
53. Migita, M.; Suzuki, T. <u>Bull. Japan. Soc. Sci. Fish.</u>, 1959,
    25, 319.
54. Takashi, R. <u>Bull. Japan. Soc. Sci. Fish.</u>, 1973, 39, 197.
55. Arai, K.; Kawamura, K.; Hayashi, C., <u>Bull. Japan. Soc. Sci.</u>
    <u>Fish.</u>, 1973, 39, 1077.
56. Finn, D. B. <u>Proc. Roy. Soc.</u> (London), 1932, B111, 396.
57. Reay, G. A.; Kuchel, C. C. <u>Gr. Brit. Dept. Sci. Ind.</u>
    <u>Research Rept. Food Invest. Board 1936</u>, 1937, 93.
58. Dyer, W. J. <u>Food Res.</u>, 1951, 16, 522.
59. Dyer, W. J.; Morton, M. L. <u>J. Fish. Res. Bd. Can.</u>, 1956,
    13, 129.
60. Dyer, W. J.; Morton, M. L.; Fraser, D. I.; Bligh, E. G.
    <u>J. Fish. Res. Bd. Can.</u>, 1956, 13, 569.
61. Connell, J. J. In "Low Temperature Biology of Foodstuffs";
    (Hawthorn, J.; Rolfe, E. J., Eds.), Pergamon Press: Oxford,
    1968, 333.
62. Dyer, W. J. In "Low Temperature Biology of Foodstuffs";
    (Hawthorn, J.; Rolfe, E. J., Eds.), Pergamon Press: Oxford,
    1968, 429.
63. Connell, J. J. In "Proteins as Human Food"; (Lowrie, R. A.,
    Ed.), Butterworths: London, 1970, 200.
64. Seagran, H. L. <u>Food Res.</u>, 1956, 23, 143.

65. Ueda, T.; Shimizu, Y.; Simidu, W. Bull. Japan. Soc. Sci. Fish., 1962, 28, 1010.
66. Noguchi, S.; Matsumoto, J. J. Bull. Japan. Soc. Sci. Fish., 1970, 36, 1078.
67. Matsumoto, J. J.; Noguchi, S. Proc. XIIIth Intern. Congr. Refrig. Washington DC, 1971, 3, 237.
68. Oguni, M.; Kubo, T.; Matsumoto, J. J. Bull. Japan. Soc. Sci. Fish., 1975, 41, 1113.
69. King, F. J., J. Food Sci., 1966, 31, 649.
70. Anderson, M. L.; Ravesi, E. M. J. Fish. Res. Bd. Can., 1969, 21, 2727.
71. Anderson, M. L.; Ravesi, E. M. J. Food Sci., 1970, 35, 199.
72. Tsuchiya, T.; Tsuchiya, Y.; Nonomura, Y.; Matsumoto, J. J. J. Biochem., 1975, 77, 853.
73. Ohnishi, M.; Tsuchiya, T.; Matsumoto, J. J. Bull. Japan. Soc. Sci. Fish., 1978, 44, 27.
74. Ohnishi, M.; Tsuchiya, T.; Matsumoto, J. J. Bull. Japan. Soc. Sci. Fish., 1978, 44, 755.
75. Jarenbäck, L.; Liljemark, A. J. Food Technol., 1975, 10, 229.
76. Jarenbäck, L.; Liljemark, A. J. Food Technol., 1975, 10, 309.
77. Ellis, D. G.; Winchester, P. M. J. Fish. Res. Bd. Can., 1958, 16, 135.
78. Connell, J. J. J. Sci. Food Agric., 1960, 11, 245.
79. Arai, K.; Takashi, R. Bull. Japan. Soc. Sci. Fish., 1973, 39, 533.
80. Ohta, F.; Yamada, T. Bull. Japan. Soc. Sci. Fish., 1978, 44, 63.
81. Partmann, W. Lebensm.-Wiss. u. Technol., 1969, 2, 124.
82. Matsumoto, J. J.; Tsuchiya, T.; Noguchi, S.; Ohnishi, M.; Akahane, T. Presented at 26th Intern. Congr. Pure and Applied Chemistry, Tokyo, Sept. 8, 1977.
83. Kawashima, T.; Arai, K.; Saito, T. Bull. Japan. Soc. Sci. Fish., 1973, 39, 207.
84. Husaini, S. A.; Alm, F. Food Res., 1955, 20, 264.
85. Connell, J. J.; Howgate, P. F. J. Food Sci., 1964, 29, 717.
86. Migita, M.; Otake, S. Bull. Japan. Soc. Sci. Fish., 1956, 22, 260.
87. Migita, M.; Otake, S. Bull. Japan. Soc. Sci. Fish., 1961, 27, 327.
88. Ueda, T.; Shimizu, Y.; Simidu, W. Bull. Japan. Soc. Sci. Fish., 1962, 28, 1005.
89. Connell, J. J. J. Sci. Food Agric., 1962, 13, 607.
90. Irisa, Y.; Ohnishi, M.; Tsuchiya, T.; Matsumoto, J. J. Presented at Annual Meeting Japanese Society of Scientific Fisheries, Tokyo, April 3, 1978.
91. Connell, J. J. Nature (London), 1959, 183, 664.
92. Slayter, H. S.; Lowey, S. Proc. Natl. Acad. Sci. U.S.A., 1967, 58, 1611.

93. Tsuchiya, Y.; Deura, K.; Tsuchiya, T.; Matsumoto, J. J. Presented at Annual Meeting Japanese Society of Scientific Fisheries, Tokyo, April 2, 1975.
94. Kimura, I.; Murozuka, T.; Arai, K. Bull. Japan. Soc. Sci. Fish., 1977, 43, 315.
95. Kimura, I.; Murozuka, T.; Arai, K. Bull. Japan. Soc. Sci. Fish., 1977, 43, 795.
96. Huxley, H. E. J. Mol. Biol., 1963, 7, 281.
97. Hanafusa, N. Low Temp. Sci., 1962, B20, 81.
98. Hanafusa, N. Low Temp. Sci., 1964, B22, 119.
99. Tsuchiya, T.; Nakamura, Y.; Ohnishi, M.; Matsumoto, J. J. Presented at Annual Meeting Japanese Society of Scientific Fisheries, Tokyo, April 2, 1977.
100. Spudich, J. A.; Watt, S. J. Biol. Chem., 1971, 246, 4866.
101. Tsuchiya, T.; Suzuki, H.; Matsumoto, J. J. Bull. Japan. Soc. Sci. Fish., 1977, 43, 1233.
102. Seki, N.; Hasegawa, E. Bull. Japan. Soc. Sci. Fish., 1978, 44, 71.
103. Love, R. H.; MacKay, E. M. J. Sci. Food Agric., 1962, 13, 200.
104. Love, R. M.; Aref, M. M.; Elerian, M. K.; Ironside, J. I. M.; MacKay, E. H.; Varela, M. G. J. Sci. Food Agric., 1965, 16, 259.
105. Tokiwa, T.; Matsumiya, H. Bull. Japan. Soc. Sci. Fish., 1969, 35, 1099.
106. Ohnishi, M.; Matsumoto, J. J. Unpublished data, 1979.
107. Love, R. M.; Robertson, I. J. Food Technol., 1968, 3, 215.
108. Love, R. M.; Yamaguchi, K.; Créac'h, Y.; Lavéty, J. Comp. Biochem. Physiol., 1976, 55B, 487.
109. Tappel, A. L. In "Cryobiology"; (Meryman, H. T., Ed.), Academic Press: London, 1966, 163.
110. Hanafusa, N. In "Water in Foods"; (Japanese Society of Scientific Fisheries, Ed.), Koseisha-Koseikaku K.K.: Tokyo, 1973, 9.
111. Shikama, K.; Yamazaki, I. Nature (London), 1961, 190, 83.
112. Shikama, K. Sci. Rept. Tohoku Univ. Ser. BIV, 1963, 27, 91.
113. Hanafusa, N. Low Temp. Sci., 1964, B24, 57.
114. Hanafusa, N. Low Temp. Sci., 1974, B32, 1.
115. Hanafusa, N. In "Freezing and Drying of Microorganisms"; (Nei, T., Ed.), University of Tokyo Press: Tokyo, 1969, 117.
116. Hanafusa, N. Refrigeration (Reito), 1973, 48, 713.
117. Gould, E. In "Technology of Fish Utilization"; (Kreuzer, R., Ed.), Fishing News (Books) Ltd.: London, 1965, 126.
118. Gould, E. J. Fish. Res. Bd. Can., 1968, 25, 1581.
119. Connell, J. J. J. Food Sci., 1966, 31, 313.
120. Tong, M. M.; Pincock, R. E. Biochemistry, 1969, 8, 908.
121. Cox, R. F.; Baust, J. G. Cryobiology, 1978, 15, 530.
122. Sakuma, R.; Aoyama, T.; Okahashi, N.; Tamiya, T.; Matsumoto, J. J. Unpublished data, 1979.
123. Wood, J. D. Can. J. Biochem. Physiol., 1959, 37, 937.

124. Tomlinson, N. J. Fish. Res. Bd. Can., 1963, 20, 1145.
125. Tomlinson, N.; Geiger, S. E. J. Fish. Res. Bd. Can., 1963, 20, 1183.
126. Wood, J. D. J. Fish. Res. Bd. Can., 1960, 16, 755.
127. Olley, J.; Lovern, J. A. J. Sci. Food Agric., 1960, 11, 644.
128. Olley, J.; Pirie, R.; Watson, H. J. Sci. Food Agric., 1962, 13, 501.
129. Lund, D. B.; Fennema, O.; Powrie, W. D. J. Food Sci., 1969, 34, 378.
130. Babbit, J. K.; Crawford, D. L.; Law, D. K. J. Agric. Food Chem., 1972, 20, 1052.
131. Sikorski, Z. E.; Kostuch, S.; Koiodzieska, I. Die Nahrung, 1975, 19, 997.
132. Fennema, O. In "Water Relations of Foods"; (Duckworth, R. B., Ed.), Academic Press: London, 1975, 539.
133. Connell, J. J. J. Sci. Food Agric., 1965, 16, 769.
134. Levitt, J. J. Theoret. Biol., 1962, 3, 355.
135. Kauzmann, W. Advan. Protein Chem., 1959, 14, 1.
136. Lapanje, S. "Physicochemical Aspects of Protein Denaturation"; John Wiley & Sons: New York, 1978, p. 331.
137. Nash, T. In "Cryobiology"; (Meryman, H. T., Ed.), Academic Press, London, 1966, 179.
138. Nishiya, K.; Takeda, F.; Tamoto, K.; Tanaka, O.; Fukumi, T.; Kitabayashi, T.; Aizawa, S. Month. Rept. Hokkaido Municpl. Fish. Exp. Station, 1961, 18, 122.
139. Nishiya, K.; Takeda, F.; Tanaka, O.; Kubo, T.; Tamoto, K. Month. Rept. Hokkaido Municpl. Fish. Exp. Station, 1961, 18, 391.
140. Matsumoto, J. J. Presented at Symposium on Fish Utilization Technology and Marketing in the IPFC Region, Manila, March 8, 1978.
141. Fukumi, T.; Tamoto, K.; Hidesato, T. Month. Rept. Hokkaido Municpl. Fish. Exp. Station, 1965, 22, 30.
142. Arai, K.; Takashi, R.; and Saito, T. Bull. Japan. Soc. Sci. Fish., 1970, 36, 232.
143. Niwa, E.; Mori, B.; Miyake, M. Bull. Japan. Soc. Sci. Fish., 1973, 39, 61.
144. Tran, V. D. J. Fish. Res. Bd. Can., 1975, 32, 1629.
145. Noguchi, S.; Matsumoto, J. J. Bull. Japan. Soc. Sci. Fish., 1971, 37, 1115.
146. Noguchi, S.; Matsumoto, J. J. Bull. Japan. Soc. Sci. Fish., 1975, 41, 243.
147. Noguchi, S.; Matsumoto, J. J. Bull. Japan. Soc. Sci. Fish., 1975, 41, 329.
148. Noguchi, S.; Shinoda, E.; Matsumoto, J. J. Bull. Japan. Soc. Sci. Fish., 1975, 41, 776.
149. Noguchi, S.; Oosawa, K.; Matsumoto, J. J. Bull. Japan. Soc. Sci. Fish., 1976, 42, 77.
150. Noguchi, S. Doctoral Thesis, Sophia University, 1974.

151. Norton, K. B.; Tressler, D. K.; Farkas, L. D. Food
     Technol., 1952, 6, 405.
152. Love, R. M.; Elerian, M. K. J. Sci. Food Agric., 1965, 16,
     65.
153. Tanikawa, E.; Akiba, M.; Shimatori, A. Food Technol., 1963,
     17, 87.
154. Love, R. M.; Abel, G. J. Food Technol., 1966, 1, 323.
155. Dyer, W. J.; Fraser, D. I. J. Fish. Res. Bd. Can., 1959,
     16, 43.
156. Bligh, E. G. J. Fish. Res. Bd. Can., 1961, 18, 143.
157. Manson, S. W. F.; Olley, J. In "The Technology of Fish
     Utilization"; (Kreuzer, R., Ed.), Fishing News (Books) Ltd.:
     London, 1965, 111.
158. Anderson, M. L.; Ravesi, E. M. J. Fish. Res. Bd. Can.,
     1958, 25, 2059.
159. Anderson, M. L.; Ravesi, E. M. J. Food Sci., 1970, 35, 551.
160. Dyer, W. J. Refrigeration (Reito), 1973, 48, 38.
161. Castell, C. H.; Smith, B.; Dyer, W. J. J. Fish. Res. Bd.
     Can., 1973, 30, 1205.
162. Dyer, W. J.; Hiltz, D. I. Environ. Canada Fish and Mar.
     Ser. News Ser., Circ. No. 45, 1974, 5.
163. Dingle, J. R.; Hines, J. A. J. Fish. Res. Bd. Can., 1974,
     32, 775.
164. Tokunaga, T. Bull. Japan. Soc. Sci. Fish., 1974, 40, 167.
165. Castell, C. H. J. Am. Oil Chem. Soc., 1971, 48, 645.
166. Connell, J. J. J. Sci. Food Agric., 1975, 26, 1925.
167. Tanaka, T. In "The Technology of Fish Utilization";
     (Kreuzer, R., Ed.), Fishing News (Books) Ltd.: London, 1965,
     121.
168. Jarenbäck, L.; Liljemark, A. J. Food Technol., 1975, 10,
     437.
169. Lovelock, J. E. Biochim. Biophys. Acta, 1953, 10, 414.
170. Lovelock, J. E. Biochim. Biophys. Acta, 1953, 11, 28.
171. Lovelock, J. E. Biochem. J., 1954, 56, 265.
172. Duerr, J. D.; Dyer, W. J. J. Fish. Res. Bd. Can., 1952, 8,
     325.
173. Snow, J. M. J. Fish. Res. Bd. Can., 1950, 7, 599.
174. Linko, R. R.; Nikkilä, O. E. J. Food Sci., 1961, 26, 606.
175. Ohta, F.; Tanaka, K. Bull. Japan. Soc. Sci. Fish., 1978,
     44, 59.
176. Ohta, F.; Yamada, T. Bull. Japan. Soc. Sci. Fish., 1978,
     44, 63.
177. Fukumi, T.; Tamoto, K.; Hidesato, T. Month. Rept. Hokkaido
     Municpl. Fish. Exp. Station, 1965, 22, 78.
178. Mazur, P. In "Cryobiology"; (Meryman, H. T., Ed.), Academic
     Press: London, 1966, 213.
179. Heber, U.; Santarius, K. A. Plant Physiol., 1964, 39, 712.
180. Karow, A. M. Jr.; Webb, W. R. Cryobiology, 1965, 2, 99.
181. Doebbler, G. F.; Rinfret, A. P. Biochim. Biophys. Acta,
     1962, 58, 449.

182. Lovelock, J. M.; Smith, A. U.   Proc. Roy. Soc. Ser. B, 1956, 145, 427.
183. Luyet, B. J.; Rapatz, G.   Biodynamica, 1958, 8, 1.
184. Meryman, H. T.   Nature (London), 1968, 218, 333.
185. Némethy, G.   In "Low Temperature Biology of Foodstuffs"; (Hawthorn, J., Ed.), Pergamon Press: Oxford, 1968, 1.
186. Frank, F.   In "Water Relations of Foods"; (Duckworth, R. B., Ed.), Academic Press: London, 1975, 3.
187. Akahane, T.; Tsuchiya, T.; Matsumoto, J. J.   Unpublished data, 1979.
188. Love, R. M.   Private communication, 1979.
189. Noguchi, S.; Matsumoto, J. J.   Bull. Japan. Soc. Sci. Fish., 1978, 44, 273.
190. Warner, D. T.   Private communication, 1979.
191. Warner, D. T.   Presented at International Symposium on Properties of Water in Relation to Food Quality and Stability, Osaka, Japan, September 1978.

RECEIVED November 8, 1979.

Preservation of Enzymes by Conjugation with Dextran

J. JOHN MARSHALL

Department of Biochemistry, School of Medicine, P.O. Box 520875,
University of Miami, Miami, FL 33152

Many enzymes are glycoprotein in nature and there is evidence to suggest that the carbohydrate in such conjugated enzymes exerts a stabilizing effect on what would otherwise be less stable proteins (1). The mechanism of stabilization by carbohydrate is not understood and it may well be that the effect of carbohydrate on stability does not represent its primary function in glycoprotein enzymes. However, the observation that carbohydrate-containing enzymes are often more stable than carbohydrate-free enzymes led the author and his colleagues to consider the possibility of stabilizing enzymes by attachment of carbohydrate to them. By this approach it was hoped to obtain modified enzymes with improved storage stability, superior activity under adverse conditions of use, resistance to the action of naturally-occurring enzyme inhibitors, and otherwise more favorable characteristics. Such tailor-made enzymes would be expected to be of value in foodstuff processing and for industrial enzyme-catalyzed conversion processes, as well as having applications as analytical and diagnostic reagents with extended shelf lives. Enzymes modified by attachment of carbohydrate might conceivably be of greater usefulness than the corresponding unmodified enzymes for medicinal purposes, including enzyme therapy of metabolic disorders.

In this article the preparation of one class of carbohydrate-enzyme conjugates, prepared by attachment of dextran to enzymes, is described in some detail and the properties of enzymes modified in this way are discussed. The molecular basis of enzyme stabilization by coupling with dextran is also considered.

Synthesis of Soluble Dextran-Enzyme Conjugates

There are many methods for covalently linking carbohydrate to enzymes, most of these having been developed for immobilization of enzymes on insoluble polysaccharide supports (2). For our work we

0-8412-0543-4/80/47-123-125$05.00/0
© 1980 American Chemical Society

selected one of the most widely used coupling methods, that involving interaction of enzymes with cyanogen bromide-activated polysaccharides (3), and adapted this procedure to make it suitable for the synthesis of soluble dextran-enzyme conjugates.

Initial efforts to activate soluble dextran with cyanogen bromide under conditions similar to those used in the case of insoluble polysaccharides (cellulose, agarose, cross-linked dextran) prior to enzyme immobilization, resulted in rapid and irreversible precipitation of the polysaccharide, presumably as a result of the cross-linking side reactions that are known to take place during cyanogen bromide activation (4). It was, therefore, necessary to develop suitable conditions for production of soluble activated dextran. The factors that are most important in determining the solubility behavior during activation are the concentrations of dextran and cyanogen bromide used in the activation reaction, and the molecular weight of the dextran being activated (5). In particular, the amount of cyanogen bromide used must be substantially lower than that used for activation of insoluble polysaccharides; amounts of cyanogen bromide greater than 0.5 gram per gram of polysaccharide almost invariably result in precipitation. An appropriate concentration of polysaccharide is about 10 mg/ml. The observation that the susceptibility to precipitation depends on the molecular weight of the dextran being activated is not surprising; while dextran of molecular weight 40,000 remains soluble on activation with cyanogen bromide used at a concentration of 0.5 gram per gram of polysaccharide, a dextran of molecular weight 2,000,000 precipitates under the same conditions. Less than 0.4 gram of cyanogen bromide per gram of polysaccharide must be used for activation in the case of the latter polysaccharide. Accordingly, we have routinely used the lower molecular weight dextran and a cyanogen bromide concentration of about 0.5 gram per gram of polysaccharide.

A conjugate of *Bacillus amyloliquefaciens* α-amylase with dextran was successfully prepared by direct addition of the α-amylase to a solution of activated dextran (6). Interaction of the enzyme with activated dextran under conditions similar to those used for synthesis of insoluble polysaccharide-enzyme conjugates (pH 9.0, 4°C for 22 hours) resulted in good retention of enzymic activity and good coupling. However attempts to prepare dextran conjugates of other enzymes (*e.g.* trypsin) in the same way resulted in rapid and extensive loss of enzymic activity during conjugation. Similar inactivation took place when the enzymes were added to a solution of activated dextran to which excess glycine had been added to block all imidocarbonate functional groupings in the activated polysaccharide. This observation indicated that enzyme inactivation was caused by by-products of the activation reaction, rather than by coupling *per se*, and pointed to the necessity of removing such by-products prior to coupling. When coupling enzymes to insoluble polysaccharides, removal of by-products of the activation reaction can be achieved simply by washing the insoluble activated carrier, but this approach is obviously unsuitable when the activated polysaccharide is soluble. The alternative methods include precipitation of the activated polysaccharide with organic solvent, gel filtration (although this manipulation might result in losses of activated polysaccharide by reaction with the gel matrix) or, preferably, a short period of dialysis. While the purification step is not always necessary as, for example, in the case of *Bacillus amyloliquefaciens* α-amylase mentioned above, dialysis of the activated polysaccharide against water for 2 hours is routinely performed before addition of the enzyme to be coupled.

Selection of conditions for the coupling reaction has represented one of the greatest problems to be overcome. It is impossible to generalize regarding the optimum coupling conditions. Initially, the conditions selected were similar to those generally employed for linking enzymes to insoluble polysaccharides, namely pH 9.0, 4°C for 22 hours, and using these conditions (hereafter referred to as standard conditions) satisfactory results were obtained with several of the enzymes we attempted to conjugate. However, a number of enzymes, including β-amylase, glucoamylase and catalase, failed to conjugate satisfactorily under such conditions, either giving unacceptably low retention of activity, or not being sufficiently well coupled. The conditions for conjugating enzymes with cyanogen bromide-activated polysaccharides were therefore investigated in detail, with particular attention to the latter three enzymes. The most important factors to be taken into consideration appear to be the duration, temperature and pH of the coupling reaction. The effect that variations in temperature and duration of coupling may have on the production of a conjugated enzyme may be illustrated by the case of β-amylase (7). Initial attempts to prepare a β-amylase-dextran conjugate under standard conditions resulted in poor retention of enzymic activity, although nearly complete coupling was obtained. A study of the time course of conjugation showed an initial rapid loss of activity, followed by a slower loss of activity later in the reaction. However, when conjugation was carried out for a short period of time to minimize activity loss, very poor coupling was achieved. When reaction was performed at a higher temperature (22°C) the initial rate of activity loss was slightly greater than the rate of activity loss at 4°C. However, the efficiency of coupling at 22°C is very much greater than at 4°C. Thus, reaction at 4°C for 4 hours resulted in 70% retention of activity but the extent of coupling was only 15%; conjugation at 22°C for the same length of time resulted in retention of 50% activity and the extent of coupling was 95%. The stability properties of dextran-enzyme conjugates are also affected by the duration and temperature of the conjugation reaction (*vide infra*).

Attachment of enzymes to activated insoluble polysaccharides is routinely carried out at pH 9.0, and this pH has been found to be suitable for synthesis of most soluble dextran-enzyme conjugates. Glucoamylase has, however, proven to be an exception. Initial attempts to conjugate the latter enzyme with cyanogen bromide-activated dextran gave results that were variable, but never satisfactory in terms of coupling efficiency. Thus, between 5 and 20% of the enzyme was usually coupled under standard conditions. Investigation of factors that might affect the efficiency of coupling showed pH to be of considerable importance, efficient conjugation only being achieved by interaction of the enzyme with activated dextran at pH values substantially lower than pH 9.0. At pH 5.0, for example, complete conjugation could be obtained. Our studies on glucoamylase-dextran conjugates are continuing, using the latter pH routinely for their synthesis. The possibility is being investigated that the low pH required for coupling of glucoamylase with activated dextran reflects that conjugation does not take place through lysine residues in the enzyme, but rather that some other amino acid side chain with a lower pK_a value than the ε-amino group of lysine (histidine?) is involved in the process. The effect of pH on the coupling of catalase to dextran is less marked but, again, better results are obtained when coupling is carried out under slightly less alkaline conditions (pH 6-8) than normally used (5). The coupling efficiency at pH values in the range 6-10 is similar, but greater retention of activity is obtained at lower pH (80% at pH 7.0) than at higher pH values (45% at pH 9.0). Below pH 6.0, the coupling efficiency decreases substantially.

Our present approach in the preparation of new dextran-enzyme conjugates is initially to test the standard conditions. If such conditions do not give satisfactory results, either in terms of retention of activity or extent of conjugation, the effect of the three important parameters, pH, temperature and duration of reaction are then investigated to establish appropriate conditions for coupling. In some cases it has also been found useful to examine the effect of these variables on the stability character-istics and other properties of the resulting dextran-enzyme conjugates.

In early studies it was observed that insolubilization of conjugated enzyme preparations tended to take place on storage at cold room temperature; in addition, lyophilization sometimes gave products that did not redissolve. It was recognized that the insolubilization was probably due to cross-link formation, and this problem has been overcome by adding to conjugation mixtures, after reaction for an appropriate length of time, excess of an amino compound (e.g. glycine) to block reactive imidocarbonate groupings that do not become involved in polysaccharide-protein linkages. When this step is included, the solubility properties of the resulting conjugated enzyme preparations, during storage or lyophilization, remain satisfactory (5).

Tests for extents of coupling are conveniently carried out by molecular-sieve chromatography of conjugated enzyme preparations on appropriate gel columns, and comparison of the elution

characteristics with those of a mixture of the corresponding free
enzyme and polysaccharide. Conjugation is indicated by elution of
the enzyme, together with dextran, at a smaller elution volume
than that of the unmodified enzyme. This procedure also serves
to remove traces of residual free enzyme prior to characterization
of the properties of a conjugated enzyme. In an effort to
simplify the process of isolating dextran-enzyme conjugates and
the measurement of extents of conjugation after reaction of
enzymes with activated dextran, we developed a process based on
the use of Concanavalin A-Sepharose (8). Enzyme-dextran conjugates
bind to the adsorbent but unmodified enzymes do not. Washing with
methyl α-D-glucoside releases conjugated enzyme. This procedure
is, of course, only applicable in cases where the enzyme being
modified is, itself, carbohydrate-free. Gel electrophoresis
under denaturing conditions (*i.e.* in the presence of sodium
dodecyl sulfate and 2-mercaptoethanol) can also be used to
determine whether coupling has taken place. Conjugated enzyme is
unable to penetrate the gel, presumably because of its high
molecular weight, whereas native enzyme migrates in the gel
according to molecular weight, as expected.

Table I shows a comparison of the conditions for coupling of
enzymes to soluble and insoluble polysaccharides.

By using the approaches described we have been able to
prepare successfully dextran conjugates of a variety of enzymes
of different types. These include α-amylase, β-amylase, glucoamylase, ribonuclease, trypsin, chymotrypsin and catalase. In
all cases we have obtained conjugated enzymes containing 50% or
more of the activity of the corresponding unmodified enzyme; in
the case of glucoamylase and ribonuclease the recovery was in the
region of 90-100%. The only enzyme we have not managed to
conjugate satisfactorily is lysozyme. While we can achieve
coupling, conjugated lysozyme preparations made under a variety
of conditions have all proven to be enzymically inactive.

Properties of Synthetic Dextran-Enzyme Conjugates

We have examined in detail the properties of the conjugated
enzymes we have synthesized. The results of carbohydrate attachment may be illustrated by considering typical properties that
are changed by the modification process, with appropriate illus-
trations from the range of conjugates we have prepared and
characterized.

Heat Stability. Most of the conjugated enzymes have been
found to have improved resistance to heat inactivation, the
magnitude of the stabilization varying from moderate to very
marked. Two amylase conjugates exhibit the greatest extent of
stabilization (10). Thus *Bacillus amyloliquefaciens* α-amylase
has a half-life of 2.5 min at 65°C; its dextran conjugate has a
half-life of 63 min under the same conditions. Sweet-potato
β-amylase has a half-life at 60°C of 5 min; its dextran conjugate

TABLE I

Preparation of Soluble and Insoluble Polysaccharide-Enzyme Conjugates

Conditions or procedure	Insoluble polysaccharide[a]	Soluble polysaccharide[b]
Typical polysaccharide concentration during activation (mg/ml)	30	10
Typical quantity of cyanogen bromide (mg/mg polysaccharide)	1-30	0.2-0.5
Removal of by-products from activated polysaccharide	wash on funnel	dialysis (alternatively organic solvent precipitation or gel chromatography)
Coupling conditions (pH, temperature, time)	pH 9.0, 4°C, 16-24 hr	must be investigated for each particular enzyme being coupled
Removal of unconjugated enzyme	wash on funnel	gel filtration chromatography or Concanavalin A-Sepharose chromatography
Test for coupling	measurement of activity associated with washed matrix	column chromatography; polyacrylamide gel electrophoresis in presence of sodium dodecyl sulfate; Concanavalin A-Sepharose chromatography

[a]Agarose (3, 9)   [b]Dextran (5)   From Reference 23

has a half-life under these conditions of 175 min (Fig. 1). It has become apparent that the degree of stabilization conferred upon an enzyme by conjugation with dextran is affected by the coupling conditions used. Thus, when β-amylase interacts with activated dextran at 22°C the stability increases with coupling time in a manner paralleling the extent of conjugation, maximum stability being achieved after coupling for 4 hours. At 4°C, maximum stability of the isolated β-amylase-dextran conjugate does not appear to be achieved even after coupling for 22 hours. The only enzyme we have not stabilized against heat inactivation is glucoamylase, itself a very stable fungal glycoenzyme, the stability characteristics of the modified enzyme being identical to those of the native enzyme.

Proteolytic Degradation. Two examples serve to illustrate the effect of polysaccharide attachment on the susceptibility of enzymes to proteolysis. The first is autolysis of trypsin and chymotrypsin; the second is degradation of ribonuclease by pepsin. Incubation of trypsin at 37°C and pH 8.1 in the presence of calcium results in autolytic digestion with a loss of 90% of its enzymic activity in 2 hours; the dextran conjugate of trypsin is essentially completely stable under these conditions (Fig. 2) (11). Similar results are found with chymotrypsin. It has long been known that ribonuclease is susceptible to inactivation by pepsin (12); the dextran conjugate of ribonuclease is, however, appreciably more stable than the native enzyme (Fig. 3). We are presently investigating the effect of the carbohydrate in the ribonuclease-dextran conjugate on the single proteolytic cleavage of this enzyme by subtilisin (13).

Removal of Cofactors. A number of enzymes are unstable in the absence of metal-ion cofactors; one such enzyme is α-amylase (14). On removal of essential calcium from α-amylase, the enzyme unfolds and is generally irreversibly inactivated, the inactivation being ascribed to cleavage of the unfolded enzyme by the traces of proteolytic enzymes that are usually present even in highly purified α-amylase preparations (14). We have examined the effect of conjugation on the inactivation of *Bacillus amyloliquefaciens* α-amylase in the presence of EDTA. The conjugated enzyme is markedly more stable than is the native enzyme. However, it is not possible to say whether the effect is due to stronger binding of calcium by the conjugated enzyme than by the native enzyme, or whether it is an effect on the proteolytic degradation step. The latter situation could arise as a result of modification of the amylase, or modification of the contaminating protease, or both. We have been unable to distinguish between these possibilities by studying the proteolysis of the amylase directly because the enzyme is resistant to exogenous proteases in the presence of calcium ions.

Effect of Protein Denaturants. Most of the conjugated enzymes we have prepared show greater resistance to inactivation than do the corresponding native enzymes when treated with protein

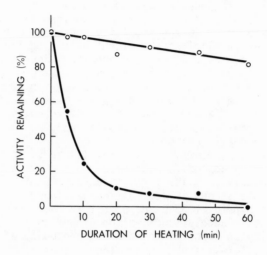

Archives of Biochemistry and Biophysics

Figure 1.   Heat inactivation (60°C) of sweet potato β-amylase (●) and its dextran conjugate (○).

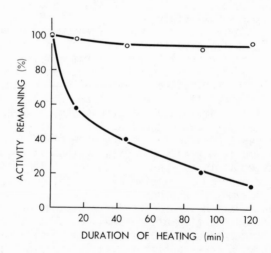

Journal of Biological Chemistry

Figure 2.   Autolysis of trypsin (●) and trypsin-dextran conjugate (○) at pH 8.1 and 37°C.

denaturants such as urea or sodium dodecyl sulfate. This situation holds in the case of proteolytic enzymes such as trypsin, where both unfolding and autolysis are likely to be involved in the denaturation process, and in the case of non-proteolytic enzymes (*e.g.* amylase, ribonuclease) where only unfolding is involved. Thus, trypsin in 8M urea loses 60% of its activity in 2 hours; trypsin-dextran conjugate loses less than 10% of its activity under the same conditions. In the presence of 8M urea and 5 mM 2-mercaptoethanol, native trypsin loses all of its activity instantaneously but the conjugated enzyme retains 60% of its activity after 2 hours under these conditions (Fig. 4).

In addition to showing improved stability in the presence of protein denaturants, we have also found dextran-conjugated enzymes to show improved activity in the presence of such agents. For example, in the absence of calcium, native trypsin is inactive in 8M urea; under the same conditions the modified enzyme displays 50% of the activity measured in the absence of urea (11).

Effect of Carbohydrate on Enzyme-Substrate Interaction. Since many of the enzymes we have conjugated with dextran have macromolecules as their natural substrates, we recognized that the conjugation process might result in unfavorable steric interactions that would impair the ability of the enzymes to interact with such substrates, in the same way that attached carbohydrate affects the susceptibility of the conjugated enzymes to proteolytic attack. We have therefore investigated the effect of carbohydrate on the interaction of conjugated enzymes with substrate.

It is not possible to generalize regarding the effect carbohydrate has on enzyme-substrate interaction; the effect varies from enzyme to enzyme. In the case of ribonuclease acting on ribonucleic acid, there is no change in the K_m value of the enzyme for ribonucleic acid after conjugation, suggesting that dextran does not interfere with the ability of ribonuclease and its substrate to combine. Similar results were found in the case of *Bacillus amyloliquefaciens* α-amylase acting on starch, and in the case of this enzyme it was possible to obtain further evidence for the lack of any effect of attached carbohydrate on enzyme-substrate interaction (6). Thus since α-amylase, during the complete degradation of starch, acts on substrate molecules of different sizes, namely polysaccharide in the early stages, megalosaccharides during the intermediate stages, and small oligosaccharides in the later stages of reaction, it might be expected that any hindrance of the enzyme to interaction with macromolecular substrate would be reflected in a different time course of hydrolysis by the native and conjugated enzymes. Such a difference was not observed, the production of reducing sugars from starch by both forms of the enzyme being identical up to 80% conversion into maltose, suggesting that both forms of the enzyme have the same relative affinity for high- and low-molecular weight substrate molecules.

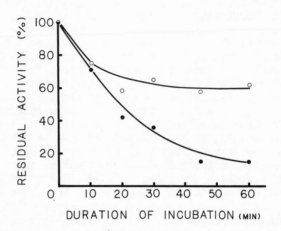

Figure 3.   Inactivation of ribonuclease (●) and ribonuclease-dextran conjugate on treatment with pepsin at pH 2.4 and 37°C.

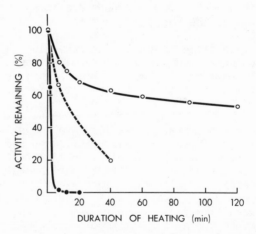

Figure 4.   Inactivation of trypsin (●) and trypsin-dextran conjugate (○) at 37°C and pH 8.1 in 8M urea and 5mM 2-mercaptoethanol. The broken line shows the rate of inactivation of trypsin-dextran conjugate after dextranase treatment.

In the case of trypsin, conjugation resulted in retention of over half the esterase activity, but the activity towards a protein substrate (casein) was essentially completely abolished (11). This is the most marked effect of carbohydrate impairing enzyme-substrate interaction that we have seen. However, it should be emphasized that preparation of the trypsin-dextran conjugate was carried out under arbitrarily chosen conditions (the standard conditions we have referred to above). The finding that we could prepare dextran conjugates of other enzymes that act on macromolecular substrates (in particular α- and β-amylases and ribonuclease) without causing such extreme loss of activity led us to believe that it should be possible to synthesize a trypsin-dextran conjugate retaining protease activity. Recent studies on the effect of changes in coupling parameters have indicated that by modification of the coupling conditions it is, indeed, possible to achieve conjugation, confer improved stability properties, and at the same time retain protease activity.

In the case of glucoamylase, we have also seen a marked effect of dextran attachment on the ability of the enzyme to interact with starch, although not with maltose. In the case of this enzyme, however, we have gone one stage further and intentionally tried to eliminate all activity towards the macromolecular substrate by suitable choice of the coupling conditions. The results of this work are described in more detail below.

Effect of Enzyme Inhibitors. Four of the enzymes we have conjugated, namely α-amylase, ribonuclease, trypsin and chymotrypsin are inhibited by naturally-occurring proteinaceous inhibitors. We have compared the effect of such inhibitors on the native and conjugated enzymes. In all cases we found resistance of the conjugated enzymes to inhibition by the respective inhibitors. Conjugated pancreatic α-amylase and conjugated ribonuclease are almost completely resistant to inhibition by phaseolamin (15) and rat liver ribonuclease inhibitor (16), respectively. Of particular interest is a comparison of the effect of several common trypsin inhibitors on native and conjugated trypsin (Table II). While native trypsin is essentially completely inhibited by all the inhibitors tested, the conjugated enzyme is inhibited to a lesser extent. The extents of inhibition of the conjugated enzyme are inversely related to the molecular weights of the inhibitors used. A more detailed study of the inhibition of trypsin and its dextran conjugate by ovomucoid (Fig. 5) showed a large proportion of the conjugated enzyme to be completely resistant to inhibition by this inhibitor, rather than all molecules being inhibited at a slower rate than in the case of the native enzyme. The resistance to inhibition can be explained in terms of steric interactions between the attached carbohydrate chains and the inhibitor molecules. Consideration of the interaction of inhibitors with conjugated

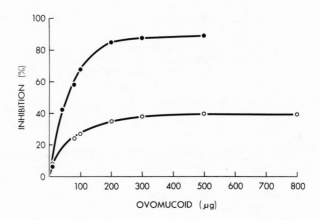

*Figure 5.   Inhibition of trypsin (●) and trypsin-dextran conjugate (○) by various
amounts of ovomucoid.*

trypsin is relevant to an understanding of the nature of the
dextran-enzyme conjugates that have been synthesized (*vide infra*).

TABLE II

Inhibition of Native Trypsin and Trypsin-Dextran

Conjugate by Protease Inhibitors[a]

	Activity remaining (%)			
Enzyme	Bovine pancreatic trypsin inhibitor	Lima bean trypsin inhibitor	Soybean trypsin inhibitor	Ovomucoid
Native	6	6	6	9
Conjugated	13	24	29	70

[a]Activities remaining after treatment of native or conjugated
trypsin (0.276 unit) with trypsin inhibitors (5 µg). From Ref. 11.

Dextranase Treatment.   Dextranase treatment of conjugated
enzymes has been shown to bring about changes in at least three
properties of conjugates, at least one of which was unexpected.
Treatment of a trypsin-dextran conjugate with dextranase
destabilized the conjugate as evidenced by its rate of inactivation in the presence of 8 M urea and 5 mM 2-mercaptoethanol after
dextranase treatment (Fig. 4).   The same treatment of the
conjugated enzyme also resulted in an increase in its
susceptibility to inhibition by ovomucoid, this increasing from
30% to 54%.   Since attachment of carbohydrate to trypsin causes
stabilization of the enzyme, it is not surprising to find that
removal of the carbohydrate results in destabilization.   The
increased susceptibility to inhibition by ovomucoid after
dextranase treatment can be explained by the removal of
unfavorable steric interactions that prevent combination of
enzyme and inhibitor.   What was surprising was the finding that
treatment of the bacterial α-amylase-dextran conjugate resulted
in an increase in its activity of about 50%.   A likely explanation
of this phenomenon is discussed below.
Specificity and Action Pattern.   In the case of some
amylolytic enzymes we have observed changes in specificity and
action pattern after conjugation with dextran.   Attempts to
conjugate glucoamylase (*vide supra*) showed substantial losses of

activity towards starch. In order to determine whether this
effect was attributable to unfavorable steric interactions
preventing efficient enzyme-substrate contact, or the result of
enzyme inactivation, we assayed the conjugated glucoamylase
against maltose and found that essentially all the activity
towards the latter substrate was retained. Having established
the factors that determine the efficiency of conjugation,
particularly the pH of the coupling reaction (*vide supra*), we
attempted to eliminate completely the activity of glucoamylase
against its macromolecular substrate, starch. Efforts to do this
involved increasing the carbohydrate:enzyme ratio, increasing the
size of the dextran to which the enzyme is coupled, and increasing
the duration of the coupling reaction. The native enzyme acts on
starch about eight times faster than it does on maltose; conjugates
prepared by modifying the coupling reaction as described act on
maltose as much as ten times faster than on starch. Thus a major
change in substrate specificity results from conjugation of
glucoamylase with dextran. While we have not yet managed to
eliminate completely activity towards starch, clearly we have
altered the specificity of the enzyme from that of a polysaccharide hydrolase towards that of an oligosaccharide hydrolase.
A conjugated glucoamylase devoid of activity towards starch could
have important practical applications, for example in the specific
hydrolysis of maltose and other small amylaceous oligosaccharides,
in the presence of polysaccharide.

    We have also, in the case of bacterial α-amylase, shown that
conjugation results in a change in action pattern. Thus, equal
activities of native and conjugated enzyme, measured in terms of
ability to release reducing sugars from starch, differ in their
effect on the iodine staining power of the substrate. The conjugated enzyme causes a lower decrease in iodine staining power
than does the native enzyme, at any given extent of hydrolysis.
The conjugated enzyme, therefore, appears to be constrained to
act at least partly in an exo-fashion, rather than in the
completely endo-fashion of the native enzyme. It remains to be
determined whether alteration of the coupling conditions will
enable us to convert an endo-acting enzyme completely into an
exo-acting enzyme.

    Miscellaneous. In addition to the effect of conjugation of
α-amylase on its heat stability, stability in denaturants, and
stability on cofactor removal, the attachment of dextran also
results in improved stability of this acid-labile enzyme at pH
values below about 5.0. Studies on the dependence of stability
of *Bacillus amyloliquefaciens* α-amylase on pH showed that the
conjugated enzyme retained 20, 15 and 7% more activity at pH 3.5,
4.0 and 4.5 than did the native enzyme.

    Many enzymes are known to bind to glass, especially when in
dilute solution. Common examples of enzymes that demonstrate
this phenomenon are β-amylase ([17]), trypsin ([18]), ribonuclease
([19]) and catalase ([20]). Conjugation of these enzymes with dextran

eliminated or substantially reduced their ability to bind to glass, presumably because of shielding of the groupings that interact with glass surfaces by the uncharged, hydrophilic carbohydrate chains attached at the surface of these enzymes.

Two "non-effects" of conjugation may also be mentioned briefly. In no case does carbohydrate attachment markedly affect the pH-activity relationship of any of the enzymes we have conjugated. This observation is not surprising since we have always conjugated enzymes with uncharged polysaccharide. It remains to be determined whether attachment of an enzyme to a charged polysaccharide affects its pH optimum.

We reasoned that coupling of polysaccharide to a chloride-dependent mammalian α-amylase might affect the dependence of the enzyme on chloride for activity. This reasoning was based on the suggestion that chloride is an allosteric effector of mammalian α-amylase, serving to convert the enzyme into an active form by interaction with lysine residues in the enzyme (21). Since conjugation was carried out in the presence of chloride, the enzyme should be conformationally locked in its most active form if the effect of conjugation is to stabilize the tertiary structure of the enzyme. However, the native and conjugated enzymes were found to show the same dependence on chloride ions for activity. This finding suggests that the conjugated enzyme still has enough conformational flexibility to require chloride ions for formation of the most catalytically-efficient conformation.

Fractionation on Sepharose. Chromatography of a trypsin-dextran-conjugate preparation on Sepharose 4B showed it to consist of a heterogenous population of molecules of very high molecular weight (Fig. 6). A comparison of the properties of trypsin-dextran conjugate "molecules" of different sizes showed little difference in properties, with the exception of the susceptibility to inhibition by ovomucoid. Highest molecular weight fractions were only slightly (25%) inhibited by this inhibitor; the lowest molecular weight molecules were inhibited to the extent of 70%.

Discussion

Nature of Soluble Dextran-Enzyme Conjugates. Column chromatography of dextran-enzyme conjugate preparations on Sepharose (Fig. 6) has shown them to consist of heterogeneous mixtures of very high molecular weight molecules. The conjugation procedure clearly does not result in formation of products containing one molecule of dextran and one molecule of enzyme. This situation arises because the amount of cyanogen bromide used in the activation step is sufficient to activate many monosaccharide residues in every dextran molecule, and each activated dextran molecule thus has the ability to cross-link polypeptide chains intermolecularly by the interaction of two or more activated monosaccharide

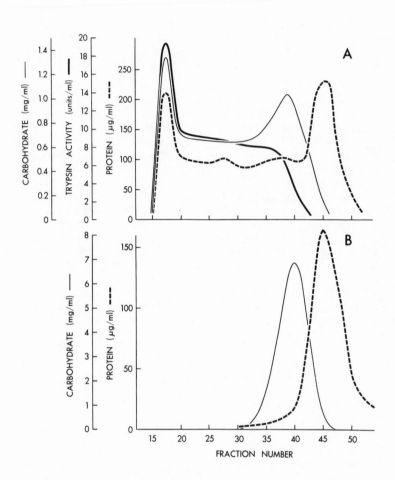

*Figure 6.   Sepharose 4B chromatography of trypsin-dextran conjugate (A). The
sample contained approximately 145 mg dextran and 14 mg of trypsin. Chromatography of a mixture containing corresponding amounts of dextran and trypsin
is shown in B.*

residues with two or more ε-amino groups of lysine in different
enzyme molecules. In this way macromolecular aggregates consisting of many enzyme molecules and many dextran molecules are produced during the coupling reaction. The enzyme molecules at the
exterior of such aggregates are likely to be fully active.
However, it is easy to envisage the activity of the enzyme
molecules in the interior of the aggregates being latent. Hence
the explanation of the apparently contradictory results obtained
when studying the bacterial α-amylase-dextran conjugate (6).
While the native and conjugated enzymes were found to have the
same K_m value for starch, the activity of the conjugated enzyme
was markedly increased by dextranase treatment. These
observations are compatible if the conjugated enzyme contains a
mixture of fully active enzyme molecules (those at the surface)
and potentially active molecules (the buried ones), the activity
of the latter being completely masked until breakdown of the
conjugated enzyme by dextranase. In a similar manner we can
explain the observations regarding the effect of trypsin
inhibitors on the trypsin-dextran conjugate, inhibitors being
unable to penetrate freely to the enzyme molecules in the interior
of the aggregates, the extents of inhibition being related to the
molecular weights of the inhibitors used, and the susceptibility
to inhibition increasing after dextranase treatment of the
conjugated enzyme (11).

Mechanism of Stabilization by Dextran. It is likely that the
stabilization conferred upon enzymes by attachment of dextran to
them arises in a number of ways. Firstly, the degree of hydration
of an enzyme molecule is probably changed by attachment of hydrophilic polysaccharide molecules to it and it is likely that heat
stability properties are affected by hydration characteristics.
Indeed, such an explanation has been given for the stabilizing
effect of the carbohydrate in naturally-occurring glycoproteins
(22). The effect of the polysaccharide in conjugated enzyme
preparations on conferring stability against proteolytic
degradation is probably two-fold. In the case of trypsin auto-
digestion, it must be recognized that a number of sites where
trypsin might act are eliminated by conjugation. Thus, since
lysine residues are involved in linkages with carbohydrate,
trypsin action is restricted to arginine residues. Secondly,
steric resistance to protease action is probably also involved,
this presumably representing the major factor preventing
chymotrypsin action on conjugated enzymes or ribonuclease
digestion by pepsin. However, the most important factor involved
in stabilization of enzymes by dextran is the effect of the
attached polysaccharide on enzyme conformation. Intramolecular
cross-linking of enzyme molecules in the macromolecular aggregates
is likely to be caused by reaction of two or more activated
monosaccharide residues in a single dextran molecule with two or
more ε-amino groupings of lysine in the *same* enzyme molecule.

The overall effect of such cross-linking on protein conformation is seen as being similar to that of disulfide bridges, this situation being apparent from the experiment where the disulfide bridges of trypsin and its dextran conjugate were reduced with 2-mercaptoethanol in the presence of 8M urea. The native enzyme lost all of its activity instantaneously; the conjugated enzyme retained 60% of its activity even after 2 hours under these conditions (Fig. 4). However cleavage of the carbohydrate bridges by dextranase treatment resulted in marked destabilization of the conjugated enzyme. Thus cross-linking by attached carbohydrate clearly plays an important part in stabilizing the conformation of enzyme molecules.

We must conclude that while we have managed by attachment of dextran to endow a number of enzymes with improved stability as we predicted, the stabilization resulting from coupling with dextran is, for the most part, due to factors that are distinct from those involved in the stabilizing role of carbohydrate in natural glycoproteins. In the latter, there is usually no cross-linking of polypeptide chains by carbohydrate; each carbohydrate moiety is attached to the polypeptide through a single linkage. Nevertheless enzymes modified by the procedure we have developed are likely to have important practical applications in biochemical technology and medicine (23).

Acknowledgments

The author is an Investigator of Howard Hughes Medical Institute. Support from the National Institutes of Health (G.M. 21258) is also acknowledged.

References

1. Pazur, J. H., Knull, H. R. and Simpson, D. L., *Biochem. Biophys. Res. Commun.* (1970) 40, 110.
2. Kennedy, J. F., *Adv. Carbohyd. Chem. Biochem.* (1974) 29, 306.
3. Axen, R. and Ernback, S., *Europ. J. Biochem.* (1971) 18, 351.
4. Kagedal, L. and Akerstrom, S., *Acta Chem. Scand.* (1971) 25, 1855.
5. Marshall, J. J. and Humphreys, J. D., *Biotechnol. Bioeng.* (1977) 19, 1739.
6. Marshall, J. J., *Carbohyd. Res.* (1976) 49, 389.
7. Marshall, J. J., *Fed. Proc., Fed. Amer. Soc. Exp. Biol.* (1976) 35, 1632.
8. Marshall, J. J. and Humphreys, J. D., *J. Chromatog.* (1977) 137, 468.
9. Cuatrecasas, P. and Anfinsen, C. B., *Methods Enzymol.* (1971) 22, 345.
10. Marshall, J. J. and Rabinowitz, M. L., *Arch. Biochem. Biophys.* (1975) 167, 777.

11. Marshall, J. J. and Rabinowitz, M. L., *J. Biol. Chem.* (1976) 251, 1081.
12. Anfinsen, C. B., *J. Biol. Chem.* (1956) 221, 405.
13. Richards, F. M., *Proc. Nat. Acad. Sci. U.S.A.* (1958) 44, 162.
14. Fischer, E. H. and Stein, E. A. *in* Boyer, P. D., Lardy, H. and Myrback, K. (eds.), *The Enzymes*, 2nd edn., Vol. 4, p. 313, Academic Press, New York, 1960.
15. Marshall, J. J. and Lauda, C. M., *J. Biol. Chem.* (1975) 250, 8030.
16. Gribnau, A. A. M., Schoenmakers, J. G., Van Kraaikamp, M., Hilak, M. and Bloemendal, H., *Biochim. Biophys. Acta* (1970) 224, 55.
17. Marshall, J. J., unpublished observations (1970).
18. Johnson, P. and Whateley, T. L., *Biochim. Biophys. Acta* (1972) 276, 323.
19. Hummel, J. P. and Anderson, B. S., *Arch. Biochem. Biophys.* (1965), 112, 443.
20. Svendsen, A., *Acta Chem. Scand.* (1953) 7, 551.
21. Levitzki, A. and Steer, M. L., *Europ. J. Biochem.* (1974) 41, 171.
22. von Euler, H. and Laurin, I., *Z. Physiol. Chem.* (1919) 108, 64.
23. Marshall, J. J., *Trends in Biochemical Sciences* (1978) 3, 79.

RECEIVED October 18, 1979.

Changes Occuring in Proteins in Alkaline Solution

JOHN R. WHITAKER

Department of Food Science and Technology, University of California, Davis, CA 95616

Proteins may be exposed to alkaline conditions during purification procedures, in the characterization of proteins, as a
step in a processing methodology or during storage. The procedure developed by Leone (1) for purification of uricase recom-
mends the treatment of the crude extract with butanol at 35° and
pH 10 for up to 18 hours. Carbohydrate chemists use alkali
treatment to distinguish between O-glycosyl linkages of carbohydrates to serine and threonine residues in proteins and amide
linkages of carbohydrates to asparagine residues in proteins (2).
Alkali treatment has also proved useful in studying structure-
function relationships of glycoproteins such as the active antifreeze glycoproteins from fishes of the Arctic and Antarctic
regions (3). In food processing, sodium hydroxide is used for
peeling of potatoes and peaches, solubilizing plant proteins,
neutralizing casein preparations (sodium caseinate), removal of
toxic constituents such as aflatoxin and protease inhibitors in
production of texturized foods and vegetable protein whipping
agents and in preparation of some special Scandinavian fish
products. Calcium hydroxide is used in processing of dough from
corn for tortillas. The most common method for obtaining protein
isolate with low nucleic acid content from microbial cells con-
sists of extracting the proteins from mechanically disrupted
cells with concentrated alkali followed by precipitation of the
extracted proteins at pH 4.5 (4-7). During storage of the egg,
the pH increases from approximately 6.5 to above 9.5 due to loss
of carbon dioxide (8).
        Adverse effects of exposing proteins to alkaline conditions
are known. As early as 1913, it was shown that severely alkali-
treated casein fed to dogs was eliminated unchanged in the feces,
that it was not attacked by putrefactive bacteria and that
trypsin or pepsin was unable to hydrolyze it (9). Ten Broeck
reported that egg albumin treated with 0.5 $\underline{N}$ NaOH for 3 weeks at
37° had no immunological properties (10). The nitrogen digesti-
bility values of 0.2 $\underline{M}$ and 0.5 $\underline{M}$ NaOH-treated casein (80°C, 1
hr), as determined in rats, was 71 and 47%, respectively, as

0-8412-0543-4/80/47-123-145$05.00/0

compared to 90% for untreated casein (11). The NPU value of
NaOH-treated soybean proteins in rats was decreased (12).
Severely alkali-treated herring meals did not support normal
growth in chicks; in fact some toxic effects were observed (13).
Dispersing soybean protein concentrates with sodium hydroxide
resulted in decreased growth in lambs (13).

Cytomegalic renal lesions and nuclear enlargement of renal
tubular cells were observed in rats fed a diet of severely
alkali-treated soybean protein (14) or a diet containing up to 3%
lysinoalanine (15). Other workers did not find such results when
less severely treated protein was fed as 20% of the total protein
and with adequate calcium supplementation (12,16) although acid
hydrolysates of the alkali-treated protein or diets containing
lysinoalanine did show effects in rats (17) similar to those
observed by Woodard and Short (14). Such effects under similar
conditions have not been observed in mice, hamsters, quails, dogs
or monkeys (18). Gould and MacGregor (19) have recently discussed some of the factors which may account for variability in
observations of the effect of feeding alkali-treated proteins.
As will be discussed later, the presence of lysinoalanine in
foods cannot be used as an indicator of alkali treatment since
lysinoalanine has been found in foods prepared without use of
alkali (20).

Reactions in Alkaline Solution

In alkaline solution, proteins are known to undergo the
following types of reactions: (a) denaturation, (b) hydrolysis
of some peptide bonds, (c) hydrolysis of amides (asparagine and
glutamine), (d) hydrolysis of arginine, (e) some destruction of
amino acids, (f) β elimination and racemization, (g) formation of
double bonds and (h) formation of new amino acids.

Denaturation. Proteins are quite susceptible to denaturation in alkaline solution because of decreased stabilization of
the tertiary structure by elimination of electrostatic interactions between carboxylate and protonated amino and guanidinium
groupings (Equations 1 and 2) and hydrogen bonding between the
hydroxyl group of tyrosine and carboxylate groups (Equation 3).

$$-COO^{\ominus}--H_3^{\oplus}N- \underset{H^+}{\overset{OH^{\ominus}}{\rightleftharpoons}} -COO^{\ominus} + -NH_2 + H_2O \qquad (1)$$

$$-COO^{\ominus}--\underset{NH_2}{\overset{NH_2}{\underset{|}{\overset{\|}{C}}}}\overset{H}{\underset{|}{N}}-R \underset{H^+}{\overset{OH^{\ominus}}{\rightleftharpoons}} -COO^{\ominus} + \underset{NH_2}{\overset{NH}{\underset{|}{\overset{\|}{C}}}}\overset{H}{\underset{|}{C}}-R + H_2O \qquad (2)$$

$$\underset{H^{\oplus}}{\overset{O}{\underset{\|}{-C}}-O^{\ominus}--H-O-\phi-} \quad \xrightarrow{OH^{\ominus}} \quad \overset{O}{\underset{\|}{-C}}-O^{\ominus} + {}^{\ominus}O-\phi- + H_2O \qquad (3)$$

Therefore, adding alkali to proteins may accomplish an increased solubilization of the protein while in the alkaline solution. However, upon adjustment to neutral or acid pH, the protein may be less soluble than originally because of denaturation. In texturization of proteins, this denaturation may be an advantage. However, in other cases such as alkali treatment to destroy aflatoxin or protease inhibitors it may be a disadvantage.

Hydrolysis. Two types of hydrolytic reactions occur in proteins at alkaline pH. These are the hydrolysis of peptide and amide bonds and the hydrolysis of arginine to ornithine. Amide bonds are hydrolyzed rapidly in alkaline solution probably as shown in Equation 4 (21).

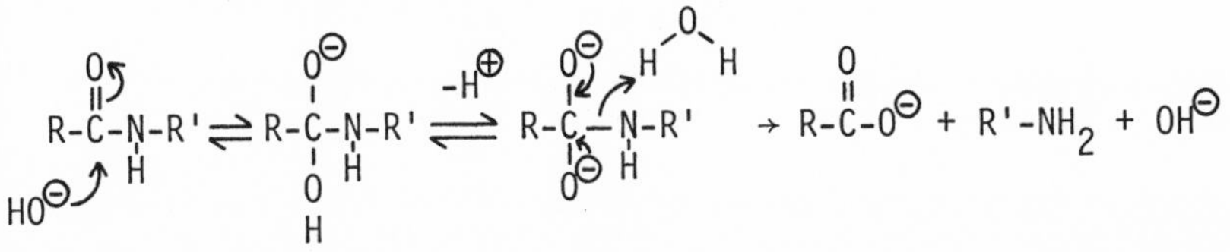

$$\underset{HO^{\ominus}}{\overset{O}{\underset{\|}{R-C}}-\underset{H}{N}-R'} \rightleftharpoons \overset{O^{\ominus}}{\underset{\underset{H}{O}}{\underset{|}{R-C}}-\underset{H}{N}-R'} \xrightarrow{-H^{\oplus}} \overset{O^{\ominus}}{\underset{\underset{H}{O^{\ominus}}}{R-C-N-R'}} \rightarrow \overset{O}{\underset{\|}{R-C}}-O^{\ominus} + R'-NH_2 + OH^{\ominus} \quad (4)$$

In this reaction, an hydroxide ion attacks the carbonyl group of the amide to form an anionic tetrahedral intermediate followed by expulsion of the -NHR' moiety. Deamidation of glutamine and asparagine residues leads to a more acidic protein derivative that may have changed solubility and functional properties. Alkali treatment also leads to loss of some of the amino acids by the processes to be described below.

In alkaline solution arginine slowly decreases and ornithine and/or citrulline is formed probably by the reaction shown in Equation 5. Ziegler et al. (22) reported that treatment of sericine with 0.1 M Na₂CO₃ at 100°C for 60 min. led to a decrease of arginine from 255 to 220 μg per gram of protein (14% change) while during the same time there was a decrease in serine from 349 to 206 μg per gram protein (41% change). Therefore, it appears that serine is less stable to alkali treatment than arginine, at least in sericine.

β Elimination and Racemization. There is some loss of the amino acids cystine, cysteine, serine, threonine, lysine and arginine during the alkaline treatment of proteins (12,22-30). Unlike arginine as shown above, loss of the other amino acids is not due to a hydrolytic reaction but rather to a β-elimination reaction (Equation 6). There is also some racemization of amino acids which can be explained by the same β-elimination reaction.

$$--HN-\underset{\underset{H}{|}}{\overset{\overset{X}{|}}{\underset{|}{CHR}}}{\overset{O}{\underset{|}{C}}}-\overset{O}{\overset{||}{C}}--\rightleftharpoons--HN-\underset{\underset{..}{|}}{\overset{\overset{X}{|}}{\underset{|}{CHR}}}{\overset{O}{\underset{|}{C}}}\overset{O}{\overset{||}{C}}--\rightleftharpoons--HN-C\overset{X}{\underset{|}{CHR}}\underset{}{\overset{\overset{\ominus}{O}}{C}}--\rightarrow--HN-\underset{|}{\overset{\overset{X}{|}}{\underset{|}{CHR}}}{\overset{O}{\underset{|}{C}}}-\overset{O}{\overset{||}{C}}-- + X^{\ominus} \quad (6)$$

(I)   (II)   (IV)   (V)

$\overset{\ominus}{OH}$

$-H^{\oplus} \| H^{\oplus}$

L-Amino acid
residue

$$--HN-\underset{\underset{X}{|}}{\overset{\overset{H}{|}}{\underset{|}{C}}}\underset{CHR}{\overset{O}{\overset{||}{C}}}-- \quad (III)$$

D-Amino acid
residue

In Equation 6, X = H, OH, O-glycosyl, O-phosphoryl, -SH,
-SCH$_2$-R, aliphatic or aromatic residue; R = H or CH$_3$.
    In the case of racemization a hydrogen can add back to the
carbanion intermediate (II) to give either the D or L amino acid
residue. Racemization of amino acids residues in proteins by
alkali treatment has been known for a long time. Dakin in 1912
(31) observed that proteins dissolved in dilute alkali underwent
a progressive increase in specific optical rotation from about
-80 to -20°. He and coworkers also observed that the amino acid
residues did not all undergo racemization at the same rate which
has been verified more recently (23,32-34). Table I shows the
relative rates of racemization of several amino acid residues in
lysozyme, phosvitin, and antifreeze glycoprotein fraction 8 as
determined by gas-liquid chromatography. It is clear the rates
of racemization vary markedly among the amino acids within a
single protein ranging from 30% for serine to 0.09% for leucine
in lysozyme and that the rates are substantially different among
proteins as shown by comparing the rates of racemization of
serine, threonine and aspartic acid in lysozyme and phosvitin.
Facilitated β elimination by having good leaving groups on the
hydroxyl of serine (phosphoserine in phosvitin) or threonine
(phosphothreonine in phosvitin and glycothreonine in the antifreeze glycoprotein) appears to lead to a smaller percentage of
racemization because of competition from the reactions leading to
compound V (Equation 6).
    The results shown in Table I are in agreement with those of
others in that the aliphatic amino acids generally show the
lowest rates of racemization. These are followed by the basic
amino acids, the aromatic amino acids, the acidic amino acids and

Table I. Racemization of Amino Acid Residues in Proteins[a,b]

Amino acid[c]	Lysozyme[d]	Phosvitin[e]	Antifreeze glycoprotein[f]
		(% D-amino acid)	
Alanine	1.02	0.89	3.56
Valine	0.42	0.00	
Leucine	0.09	1.20	
Allo-Isoleucine	0.50	0.45	
Phenylalanine	2.98	3.91	
Tyrosine	2.62	-	
Proline	0.68	1.05	0.06
Serine	30.1	2.47	
Allo-Threonine	12.0	5.52	4.87
Aspartic acid	16.2	6.82	
Glutamic acid	2.82	1.33	
Lysine	0.96	2.61	

[a]Ref. 34.

[b]Determined by gas chromatography as described in Reference 35.

[c]Determined after hydrolysis of the alkali treated sample in 6 N HCl for 22 hours at 110°C.

[d]3.3 mg/ml lysozyme in 0.5 N NaOH for 2.5 hours at 22°C.

[e]3.8 mg/ml phosvitin in 0.123 N NaOH for 30 min. at 60°C.

[f]Antifreeze glycoprotein fraction 8 at 5 mg/ml in 0.5 N NaOH for 21 hrs at 22°C.

the aliphatic hydroxy amino acids having the fastest rate of β
elimination. Amino acid residues undergo more rapid racemization
in proteins than when free. This is because the electron density
of the amino and carboxylate groups of the free amino acid in the
vicinity of the α-carbon decreases the attack by the hydroxide
ion.

   Not all elimination of hydrogen from the α-carbon of an
amino acid residue leads to racemization. In the case of serine,
threonine, cystine and cysteine intermediate II can continue via
the pathway of intermediate IV to give the dehydroamino acid (V).
The requirement for this pathway is that X be a good leaving
group such as -OH, -SH and -S-CH$_2$-R. Attachment of phosphoryl
and glycosyl groups lead to even faster rates. In alkali solution one might expect that phosphate groups would be removed from
0-phosphoserine-containing proteins by hydrolysis and by β
elimination. With phosvitin it was found that greater than 85%
of the phosphate group was removed by β elimination as measured
by increase in absorbance at 241 nm due to formation of dehydroalanine (36). Anderson and Kelley (37) had postulated as early
as 1959 that the mechanism of β elimination of the phosphate
group in alkaline solution would follow the general mechanism
outlined in Equation 6.

   The effect of alkali on the degradation of cystine has been
studied extensively in both model systems as well as in proteins.
Three models have been proposed to explain the degradation.
These are: (a) β elimination, (b) α elimination and (c) hydrolysis. In β elimination, the proposed reactions are shown
in Equation 7. Therefore, the stoichiometry of the reaction is
two moles of dehydroalanine, one mole of elemental sulfur and one
mole of disulfide as shown by the overall reaction (Equation 8).

$$\overset{\backprime}{H}C\text{-}CH_2\text{-}S\text{-}S\text{-}CH_2\text{-}C\overset{\prime}{H} + 2\ \overset{\ominus}{O}H \rightarrow 2\ \overset{\backprime}{H}C\text{=}CH_2 + S^{\ominus} + S + 2\ H_2O \qquad (8)$$

Friedman (38) has proposed that the two hydrogens on the α-car-
bons may be eliminated simultaneously thus leading directly to
the same final products.

   In the α-elimination mechanism (39) as shown in Equation 9,
hydrogen extraction by the base alpha to the sulfur results in
formation of a carbanion which may rearrange via route $\underline{a}$ or $\underline{b,c}$
to give cysteine and a thioaldehyde which decomposes in alkaline
solution to an aldehyde and hydrogen sulfide. Thus, the stoichiometry of this reaction is (Equation 10):

$$\overset{\backprime}{H}C\text{-}CH_2\text{-}S\text{-}S\text{-}CH_2\text{-}C\overset{\prime}{H} + 2\ \overset{\ominus}{O}H \rightarrow \overset{\backprime}{H}C\text{-}CH_2\text{-}S^{\ominus} + \overset{\backprime}{H}C\text{-}CHO + HS^{\ominus} + H_2O \qquad (10)$$

Therefore, the α-elimination mechanism alone could not lead to
formation of dehydroalanine.

The proposed hydrolysis mechanism (40) shown in Equation 11 leads to a final overall stoichiometry of (Equation 12):

$$2\overset{\ominus}{H}C\text{-}CH_2\text{-}S\text{-}S\text{-}CH_2\text{-}C\overset{\ominus}{H} + 4\overset{\ominus}{O}H \rightarrow 3\overset{\ominus}{H}C\text{-}CH_2\text{-}S^{\ominus} + \overset{\ominus}{H}C\text{-}CH_2\text{-}SO_2^{\ominus} + 2H_2O \quad (12)$$

The second molecule of cysteine and a sulfinic acid result from the disproportionation of two molecules of a sulfenic acid (Equation 11).

Nashef et al. (41) have carefully studied the stoichiometry of products formed from alkali treatment of lysozyme and α-lactalbumin. The stoichiometry was consistent with the β-elimination mechanism and could not be explained by either the α-elimination or the hydrolysis mechanisms. Neither the α-elimination nor hydrolysis mechanism alone will explain the formation of dehydroalanine and the subsequent addition products that are observed (see below). The β-elimination mechanism is also consistent with the products found when keratin is treated in an alkaline solution containing ^{35}S-sulfide (42). The observation that cystine as the free amino acid cannot undergo lanthionine formation also favors the β-elimination mechanism (43). In the free amino acid, the amino group and the carboxylate anion attached to the α-carbon atom (see Equations 7, 9, 11) generate a high electron density and prevent the abstraction of the α-hydrogen atom by base. In the presence of calcium or strontium hydroxide, lanthionine is formed from the free cysteine indicating that the calcium or strontium ion, by complexation, can reduce the electron charge about the α-carbon atom (42).

The β-elimination reaction (Equation 6) is sensitive to pH, temperature and presence of other ions. Table II shows the effect of hydroxide ion concentration on the initial rate of β elimination of phosphate from phosphoserine in phosvitin (36) and glycosyl groups from threonine in antifreeze glycoproteins (44). The initial rate is directly proportional to hydroxide ion concentration over the range investigated. The β elimination of glycosyl groups from threonine in antifreeze glycoprotein-8 is some 12 times faster at 60°C than the rate of β elimination of phosphate groups from phosphoserine in phosvitin (Table II).

Nashef et al. (41) also reported that the rate of β elimination from cystine was directly dependent on hydroxide ion concentration although the relationship was not linear perhaps because of the complexity of the reaction (Equation 7). Sternberg and Kim (20) found the rate of lysinoalanine formation in casein to be dependent on hydroxide ion concentration. Touloupais and Vassiliadis (45) also found the rate of lysinoalanine formation in wool to be pH dependent. These workers did not measure the rate of β elimination, therefore the rate determining step is not known. These results on proteins appear to be in contradiction to those of Samuel and Silver (46) who reported that hydroxide ion concentration had no effect on the rate of β elimination from free phosphoserine between pH 7 and 13.5. Because of the effect

Table II. Effect of Hydroxide Ion Concentration and
Calcium Ion on Initial Rate of β Elimination
of Phosvitin and Antifreeze Glycoprotein-8 (AFGP-8)

Protein	$[OH^-]$ $(\underline{M} \times 10^2)$	Initial rate/$[OH^-]$ $(\underline{M}^{-1} min^{-1})$
AFGP-8[a]	0.001	4.13[b]
	0.01	0.614
	0.1	0.570
	1.0	0.226
	50	0.408
		ave = 0.454 (0.647 at 60°C)[c]
Phosvitin[d]	1.74	0.0370
	5.41	0.0514
	12.3	0.0622 (0.934)[e]
	18.7	0.0647
		ave = 0.0538

[a]Reference 44. Performed at 50.0°C in 0.2 $\underline{M}$ phosphate-NaOH
buffers and at 3.70 X 10^{-5} $\underline{M}$ AFGP-8.

[b]Left out of average.

[c]Calculated using E_a = 9.60 kcal/mol.

[d]Reference 36. Performed at 60.0°C in KCl-NaOH buffers with
1.0 X 10^{-6} to 1.1 X 10^{-5} $\underline{M}$ phosvitin. Rates corrected to
ionic strength of 0.170.

[e]As in d except in presence of 1.12 m$\underline{M}$ $CaCl_2$.

Table III. Effect of Temperature on the β Elimination of O-Phosphoserine, O-Glycothreonine and Cystine Groups in Proteins

Protein	E_a (kcal/mol)	$\Delta H^{\ddagger}$ (kcal/mol)	$\Delta G^{\ddagger a}$ (kcal/mol)	$\Delta S^{\ddagger a}$ (cal/mol/deg)
AFGP-8[b] (O-glycothreonine)	9.60	8.94	22.4	-39.9
Phosvitin[c] (O-phosphoserine)	20.2	19.5	24.1	-13.8
with $CaCl_2$[d]	20.8	20.1	22.5	-7.21
Lysozyme[e] (cystine)	23.8	23.1	20.2	8.71
GAX ovomucoid[f] (cystine)	14.2	13.5	20.3	-20.4

[a] At 60.0°C.

[b] Reference 44. Antifreeze glycoprotein-8 in 0.5 $\underline{N}$ NaOH at 2.22 X 10^{-3} $\underline{M}$ protein concentration.

[c] Reference 36. At 0.170 ionic strength with phosvitin concentrations of 1.0 - 11 X 10^{-6} $\underline{M}$ phosvitin.

[d] Reference 36. As in c except 7.47 X 10^{-4} $\underline{M}$ $CaCl_2$ added.

[e] Reference 41. Reactions were in 0.1 $\underline{N}$ NaOH at 1 X 10^{-5} $\underline{M}$ lysozyme.

[f] Reference 41. Reactions were in 0.1 $\underline{N}$ NaOH at 1 X 10^{-5} $\underline{M}$ golden pheasant ovomucoid (GAX ovomucoid).

of the electron density of the amino and carboxylate groups on
β elimination (see above) in free phosphoserine the results
are not directly comparable.

β Elimination from serine, threonine and cystine is temperature dependent as shown by the data of Table III. The effect of
temperature ($\Delta H^{\ddagger}$) on the initial rates of β elimination is in the
increasing order of removal of glycosyl groups from glycothre-
onine in antifreeze glycoprotein, sulfur from cystine in GAX
ovomucoid, phosphoryl groups from phosphoserine in phosvitin and
sulfur from cystine in lysozyme. Therefore, the environment in
the proteins may have more effect on the influence of temperature
on the rate than the type of group undergoing β elimination
(compare $\Delta H^{\ddagger}$ of 13.5 and 23.1 kcal/mol for GAX ovomucoid and
lysozyme, respectively in Table III).

The rate of β elimination is also influenced by the type of
ions present in the solution. Sen et al. (36) showed that the
rate of β elimination of phosphoserine in phosvitin was markedly
enhanced by the addition of calcium chloride (Table IV).
Touloupis and Vassiliadis (45) reported that sodium phosphate
enhanced the rate of formation of lysinoalanine in wool several
fold over that found in sodium carbonate solutions of equal pH.
It is likely that the observed effect was on the rate of β
elimination rather than the addition reaction. As reported

Table IV. Effect of Calcium Chloride Concentration on
Initial Rates of β Elimination of Phosvitin and
Addition of Lysine to the Dehydroalanine Formed[a]

$CaCl_2$ ($\underline{M}$ X 10^4)	Initial rate of β-elimination ($\underline{M}^{-1}$ min^{-1} X 10^2)	Initial rate of addition ($\underline{M}^{-1}$ min X 10^2)
0	4.74	2.5
3.36	24.8	1.4
5.60	43.2	1.9
8.96	53.8	4.3
11.2	93.4	6.4

[a]Reference 36. At 60°C, 0.123 $\underline{N}$ NaOH, 5.54 X 10^{-6} $\underline{M}$ phosvitin.

above, Feairheller et al. (42) found that calcium and strontium ions permitted the formation of lanthionine from cystine in
free form presumably by reduction of the high electron density in
the neighborhood of the α-carbon. We suspect the same general
effect is operative in the case of calcium ion on the rate of β
elimination of phosphoserine in phosvitin. The specific effect
is to mask the negative charges on the phosphate group thus
permitting the hydroxide ion to abstract more easily the hydrogen

from the α-carbon (Equation 6).

<u>Addition Reaction</u>. The double bond of dehydroalanine and β-methyldehydroalanine formed by the β-elimination reaction (Equation 6) is very reactive with nucleophiles in the solution. These may be added nucleophiles such as sulfite (<u>44</u>), sulfide (<u>42</u>), cysteine and other sulfhydryl compounds (<u>20,47</u>), amines such as α-N-acetyl lysine (<u>47</u>) or ammonia (<u>48</u>). Or the nucleophiles may be contributed by the side chains of amino acid residues, such as lysine, cysteine, histidine or tryptophan, in the protein undergoing reaction in alkaline solution. Some of these reactions are shown in Figure 1. Friedman (<u>38</u>) has postulated a number of additional compounds, including stereo-isomers for those shown in Figure 1, as well as those compounds formed from the reaction of β-methyldehydroalanine (from β elimination of threonine). He has also suggested a systematic nomenclature for these new amino acid derivatives (<u>38</u>). As pointed out by Friedman the stereochemistry can be complicated because of the number of asymmetric carbon atoms (two to three depending on derivative) possible.

Addition to the double bond of dehydroalanine (or β-methyldehydroalanine) involves nucleophilic attack by compounds containing S, O or NH as shown by Figure 1. The overall reaction may be written as shown in Equation 13

$$--HN-\underset{\underset{CHR}{\|}}{C}-CO-- \quad + \quad R'-X-H \rightarrow \quad --HN-\underset{\underset{XCHR}{|}}{\overset{\overset{H}{|}}{C}}-CO-- \tag{13}$$

where R is H or CH$_3$ and R' may represent the protein, the remainder of the side chain of an amino acid derivative or H as in H$_2$S.

The rate of addition of protein-bound lysine to dehydroalanine has been shown to be pH dependent and temperature dependent but relatively independent of ionic strength and calcium chloride concentration (<u>36</u>). Because of the nucleophilic nature of the addition reaction, it is not surprising that the initial rate of addition should be pH dependent until all the nucleophile is in the correct form (unprotonated ε-amino group in the case of lysine, pK$_a$ ∿ 10.5). The effect of temperature on the rate of the addition reaction to dehydroalanine in phosvitin is shown in Table V. It is interesting that, although CaCl$_2$ appears not to affect the rate of the addition reaction at 60°C (Table IV), it has a rather marked effect on ΔS‡. Feairheller et al. (<u>42</u>) found that added calcium and strontium hydroxides permitted lanthionine formation from free cystine. Touloupis and Vassiliadis (<u>45</u>) reported that the rate of formation of lysinoalanine in wool was faster in sodium phosphate than sodium carbonate at the same pH and temperature. While the rate of β elimination of phospho-

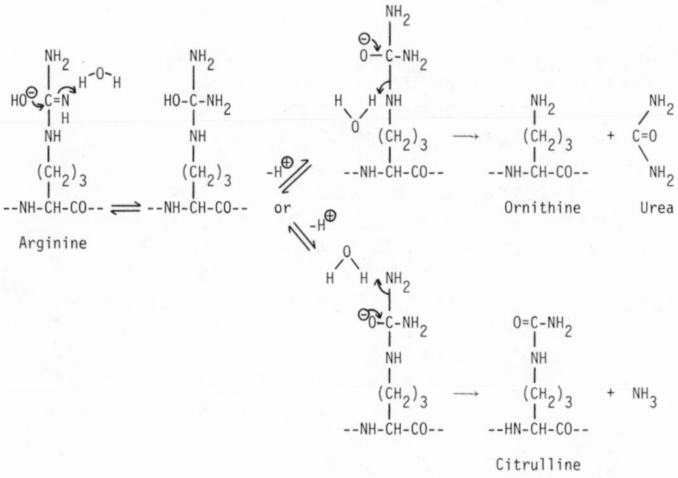

Figure 1.   New amino acids which may be formed through reaction of a dehydroalanine residue with internal or external nucleophiles in alkali treated proteins.

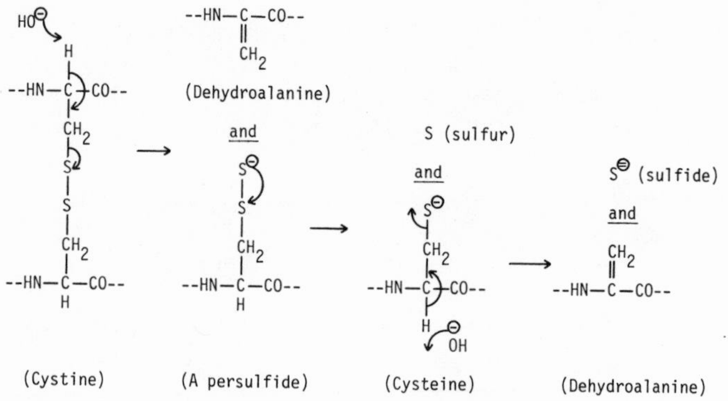

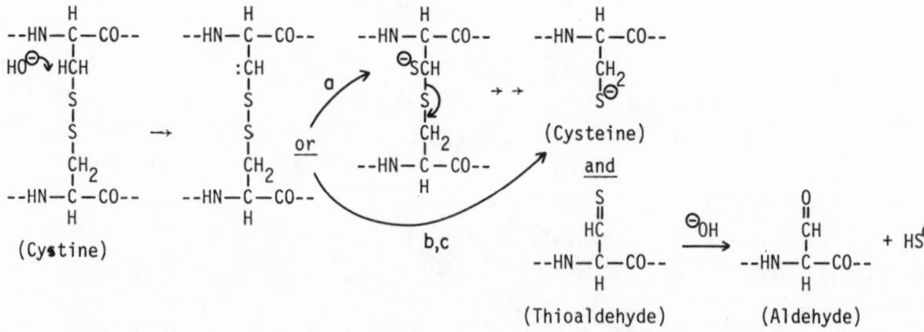

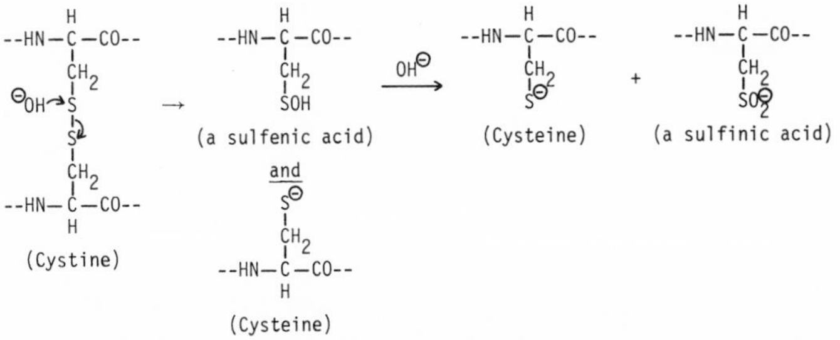

Equation (11):

--HN—C(H)—CO--, CH$_2$, S with $^\ominus$OH (Cystine) → --HN—C(H)—CO--, CH$_2$, SOH (a sulfenic acid) and --HN—C(H)—CO--, CH$_2$, S$^\ominus$ (Cysteine) $\xrightarrow{OH^\ominus}$ --HN—C(H)—CO--, CH$_2$, S$^\ominus$ (Cysteine) + --HN—C(H)—CO--, CH$_2$, SO$_2^\ominus$ (a sulfinic acid)   (11)

Lysinoalanine
HOOC(H$_2$N)CH-CH$_2$-NH-(CH$_2$)$_4$-CH(COOH)(NH$_2$)
(Bohak, 1964; Ref. 50)

3-T-Histidinoalanine
(postulated by Finley and Friedman, 1977; Ref. 51)

Unknown
(postulated by Finley and Friedman, 1977; Ref. 51)

β-Aminoalanine
HOOC(H$_2$N)CH-CH$_2$-NH$_2$
(Asquith and Skinner, 1970; Ref. 49)

Arginine derivative
HN(CH$_2$)$_3$CH(NH$_2$)-COOH
(postulated by Finley and Friedman, 1977; Ref. 51)

Dehydroalanine
HOOC(H$_2$N)C=CH$_2$

Ornithinoalanine
HOOC(H$_2$N)CH-CH$_2$-NH-(CH$_2$)$_3$-CH(COOH)(NH$_2$)
(Ziegler et al., 1967; Ref. 22)

Lanthionine
HOOC(H$_2$N)CH-CH$_2$-S-CH$_2$-CH(COOH)(NH$_2$)
(Horn et al., 1941; Ref. 52)

Lysine, Arginine, Histidine, Tryptophan, Ornithine, Cysteine, NH$_3$

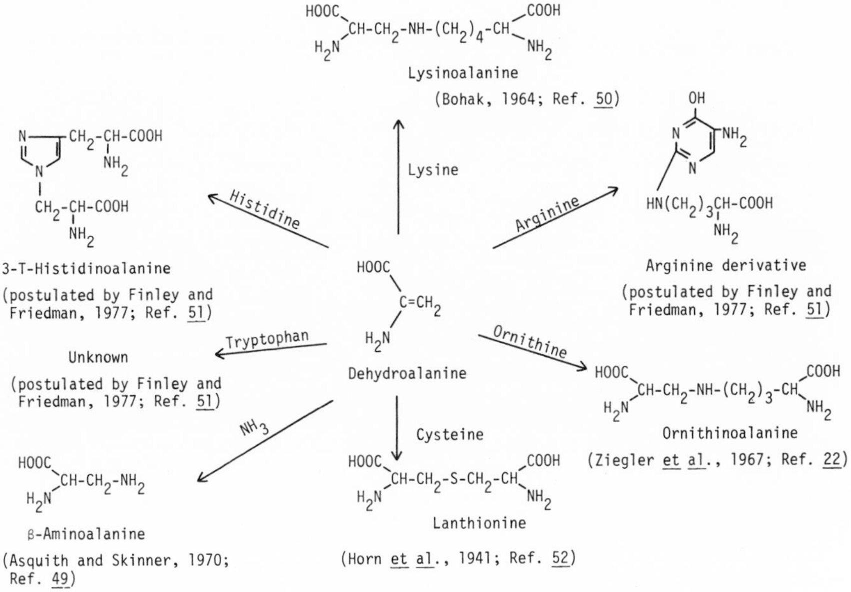

Figure 1. Continued

Table V. Effect of Temperature on the Addition of
Internal Lysine to β Eliminated Phosvitin[a]

Reaction	E_a (kcal/mol)	$\Delta H^{\ddagger}$ (kcal/mol)	$\Delta G^{\ddagger b}$ (kcal/mol)	$\Delta S^{\ddagger b}$ (cal/mol/deg)
No CaCl$_2$	25.0	24.3	24.6	-0.90
CaCl$_2$[c]	27.6	26.9	24.2	+8.11

[a]Reference 36. Rate data corrected to ionic strength of 0.170.
[b]at 60.0°C.
[c]7.47 X 10^{-4} $\underline{M}$ CaCl$_2$.

serine in phosvitin was quite dependent on ionic strength, the
addition reaction was completely independent of ionic strength
(36).
      The rate of nucleophile addition to the double bond is
dependent on the nature of the nucleophile as would be expected.
Finley et al. (47) have measured the relative effectiveness of
the sulfhydryl group of L-cysteine and the ε-amino group of
α-N-acetyl-L-lysine in adding to the double bond of N-acetyl
dehydroalanine. At equal concentrations of the reactive species,
cysteine adds some 31 times more rapidly to the double bond than
does α-N-acetyl-L-lysine (47). However, when one compares these
two compounds at the same pH the relative rates in favor of
cysteine (pK of sulfhydryl group = 8.15) versus α-N-acetyl-L-
lysine (pK of ε-amino group = 10.53) are most impressive at lower
pH's (Table VI). Therefore, it has been recommended that cys-

Table VI. Relative Rates of Addition of the Sulfhydryl Group
of Cysteine and the ε-Amino Group of α-N-Acetyl-L-Lysine to the
Double Bond of N-Acetyldehydroalanine at Different pH Values[d]

pH	Relative rate Cysteine SH/Lysine ε-NH$_2$
7.0	5000
8.0	2300
9.0	410
10.0	133
11.0	43
12.0	34

[a]Adapted from Reference 47.

teine be added to foods during alkali processing in order to
minimize lysinoalanine formation (20,47) and prevent loss of the
essential amino acid lysine.

Significance and application of alkali treatment of proteins

The reactions of proteins in alkaline solution are very important from a number of standpoints. We have already discussed
several uses of alkali treatment in food processing in the in-
troduction. When contact between the food and alkali is kept to
a minimum at the lowest temperature possible with adequate control of mixing, etc. there is presently no apparent reason to
discontinue its use. Low levels of lysinoalanine occur in food
which has been processed in the absence of added alkali, even at
pH 6 and in the dry state (20). For example, the egg white of an
egg boiled three minutes contained 140 ppm of lysinoalanine while
dried egg white powder contained from 160 to 1820 ppm of lysinoalanine depending on the manufacturer (20). No lysinoalanine was
found in fresh egg white. β Elimination and addition of lysine
to the double bond of dehydroalanine reduce the level of the
essential amino acid lysine. This can be prevented by adding
other nucleophiles such as cysteine to the reaction. Whether
lysinoalanine (and other compounds formed by addition reactions)
is toxic at low levels in humans is not known.

The β-elimination and addition reactions may be important in
the texturizing of foods extruded from alkaline solution. Should
this prove to be required in texturization other means of forming
crosslinkages between protein molecules should be developed. In
this connection, β elimination in the presence of dithiol compounds may prove useful. Following β elimination and addition of
the dithiol compound to the double bond via one of the sulfhydryl
groups, the modified protein could then be allowed to oxidize in
air to form disulfide bridges. The physical and digestibility
properties of such a protein would be most interesting. Hydrogen
sulfide, which adds to the double bond (42), could also be used
for this purpose.

The β-elimination reaction could also be used to change the
solubility properties of a protein. For example, alkali treatment in the presence of sodium sulfite leads to incorporation of
sulfonate groups into the protein (44,53) which would increase
its water solubility and probably change its functional properties.

The β-elimination reaction is used routinely to distinguish
O-glycosyl linkages of carbohydrate to serine and threonine in
proteins from amide linkages of carbohydrates to asparagine
residues in proteins (2). In alkali, the O-glycosyl groups
undergo β elimination to form dehydroalanine (from serine) and
β-methyldehydroalanine (from threonine) while the amide-linked
carbohydrate is not removed by such treatment. The β-elimination
reaction has been used to show the essentiality of the carbohy-

drate side chains for the freezing-point depressing activity and lectin-inhibiting properties of the antifreeze glycoproteins (3).

β Elimination has been used to show differences among the disulfide bonds in various proteins including the ovomucoids (54). The β-elimination reaction has also been used to replace the hydroxyl group of the essential serine residue of subtilisin with a sulfhydryl group (55). The thiolsubtilisin had a small fraction of the activity of subtilisin but it has been quite useful in mechanistic studies of the serine and sulfhydryl proteases.

As a consequence of dehydroalanine and β-methyldehydro-alanine formation specific bond cleavage can occur. Ebert et al. (56) have shown that addition of cysteine to the double bonds of polydehydroalanine and copolymers of dehydroalanine results in increased solubility and decrease in molecular weight because of peptide bond cleavage caused by formation of a thiazolidine. This reaction can be used for selective peptide chain cleavage of cysteine-containing polypeptides and proteins under rather mild conditions. Mild acid treatment of dehydroalanine-containing polypeptides and proteins leads to specific peptide bond cleavage with formation of pyruvate and ammonia (57-59)).

The difference in spectral properties of tyrosine in neutral vs alkaline solution can be used to determine the tyrosine content of proteins and by difference tryptophan (60).

Acknowledgments

The author thanks Vicky Crampton for checking the references and Clara Robison for typing of the manuscript.

Literature Cited

1. Leone, E. Biochem. J., 1953, 54, 393.
2. Downs, F.; Pigman, W. Meth. Carbohydr. Chem., 1976, 7, 200.
3. Ahmed, A. I.; Osuga, D. T.; Feeney, R. E. J. Biol. Chem., 1973, 248, 8524.
4. Cunningham, S. D.; Cater, C. M.; Matill, K. F. J. Food Sci., 1975, 40, 732.
5. Vananuvat, P.; Kinsella, J. E. J. Agric. Food Chem., 1975, 23, 216.
6. Lindblom, M. A. Lebensmittel.-Wiss U-Technol., 1974, 7, 295.
7. Lindblom, M. A. Biotechnol. Bioeng., 1974, 16, 1495.
8. Feeney, R. E.; Allison, R. G. "Evolutionary Biochemistry of Proteins. Homologous and Analogous Proteins from Avian Egg Whites, Blood Sera, Milk, and Other Substances"; John Wiley and Sons: New York, N.Y., 1969.
9. Dakin, H. D.; Dudley, H. W. J. Biol. Chem. 1913, 15, 271.
10. Ten Broeck, C. J. Biol. Chem. 1914, 17, 369.
11. Cheftel, C.; Cuq, J. L.; Provansal, M.; Besancon, P. Rev.

Fr. Corps Gras., 1976, 1, 7.

12. de Groot, A. P.; Slump, P. J. Nutr., 1969, 98, 45.

13. Cheftel, C. In "Food Proteins"; (Whitaker, J. R.; Tannenbaum, S. R., Eds.), Avi: Westport, Conn., 1977, p. 401.

14. Woodard, J. C.; Short, D. D. J. Nutr., 1973, 103, 569.

15. Woodard, J. C.; Short, D. D.; Alverez, M. R.; Reyniers, J. In "Protein Nutritional Quality of Foods and Feeds"; Part 2 (Friedman, M., Ed.), Marcel Dekker, Inc.: New York, 1975, p. 595.

16. Van Beek, L.; Feron, V. J.; de Groot, A. P. J. Nutr., 1974, 104, 1630.

17. de Groot, A. P.; Slump, P.; Van Beek, L.; Feron, V. J. Abstr., 35th Annual Meeting of the Institute of Food Technologists, Chicago, 1975.

18. de Groot, A. P.; Slump, P.; Feron, V. J.; Van Beek, L. J. Nutr., 1976, 106, 1527.

19. Gould, D. H.; MacGregor, J. T. In "Protein Crosslinking. Nutritional and Medical Consequences"; (Friedman, M., Ed.), Plenum Press: New York, Adv. Expt'l. Med. Biol., 1977, 86B, 29.

20. Sternberg, M.; Kim, C. Y. In "Protein Crosslinking. Nutritional and Medical Consequences"; (Friedman, M., Ed.), Plenum Press: New York, Adv. Expt'l. Med. Biol., 1977, 86B, 73.

21. Jencks, W. P. "Catalysis in Chemistry and Enzymology"; McGraw-Hill Book Co.: New York, 1969, p. 523.

22. Ziegler, K. L.; Melchert, I.; Lürken, C. Nature (London), 1967, 214, 404.

23. Pickering, B. T.; Li, C. H. Arch. Biochem. Biophys., 1964, 104, 119.

24. Geschwind, I. I.; Li, C. H. Arch. Biochem. Biophys., 1964, 106, 200.

25. Mellet, P. Text. Res. J., 1968, 38, 977.

26. Blackburn, S. "Amino Acid Determination, Method and Techniques"; Marcel Dekker, Inc.: New York, 1968.

27. Parisot, A.; Derminot, J. Bull. Inst. Text. Fr., 1970, 24, 603.

28. Whiting, A. H. Biochim. Biophys. Acta, 1971, 243, 332.

29. Gottschalk, A. "Glycoproteins"; Elsevier Publ. Co.: New York, 1972.

30. Provansal, M. M. P.; Cuq, J.-L. A.; Cheftel, J.-C. J. Agric. Food Chem., 1975, 23, 938.

31. Dakin, H. D. J. Biol. Chem., 1912, 13, 357.

32. Levene, P. A.; Bass, L. W. J. Biol. Chem., 1928, 78, 145.

33. Tannenbaum, S. R.; Ahern, M.; Bates, R. P. Food Technol. (Chicago), 1970, 24, 604.

34. Pollock, G. E.; Feeney, R. E.; Whitaker, J. R. Unpublished data, 1979.

35. Pollock, G. E.; Cheng, C.-N.; Cronin, S. E. Anal. Chem.,

1977, 49, 2.
36. Sen, L. C.; Gonzalez-Flores, E.; Feeney, R. E.; Whitaker, J. R. J. Agric. Food Chem., 1977, 25, 632.
37. Anderson, L.; Kelley, J. J. J. Am. Chem. Soc., 1959, 81, 2275.
38. Friedman, M. In "Protein Crosslinking. Nutritional and Medical Consequences"; (Friedman, M., Ed.), Plenum Press: New York, Adv. Exptl. Med. Biol., 1977, 86B, 1.
39. Danehy, J. P.; Elia, V. J. J. Org. Chem., 1971, 36, 1394.
40. Schiller, R.; Otto, R. Chem. Ber., 1876, 9, 1637.
41. Nashef, A. S.; Osuga, D. T.; Lee, H. S.; Ahmed, A. I.; Whitaker, J. R.; Feeney, R. E. J. Agric. Food Chem., 1977, 25, 245.
42. Feairheller, S. H.; Taylor, M. M.; Bailey, D. G. In "Protein Crosslinking. Nutritional and Medical Consequences"; (Friedman, M., Ed.), Plenum Press: New York, Adv. Exptl. Med. Biol., 1977, 86B, 177.
43. Danehy, J. P. Int. J. Sulfur Chem. B., 1971, 6, 103.
44. Lee, H. S.; Osuga, D. T.; Nashef, A. S.; Ahmed, A. I.; Whitaker, J. R.; Feeney, R. E. J. Agric. Food Chem., 1977, 25, 1153.
45. Touloupis, C.; Vassiliadis, A. In "Protein Crosslinking. Nutritional and Medical Consequences"; (Friedman, M., Ed.), Plenum Press: New York, Adv. Exptl. Med. Biol., 1977, 86B, 187.
46. Samuel, D.; Silver, B. L. J. Chem. Soc., 1963, 289.
47. Finley, J. W.; Snow, J. T.; Johnston, P. H.; Friedman, M. In "Protein Crosslinking. Nutritional and Medical Consequences"; (Friedman, M., Ed.), Plenum Press: New York, Adv. Exptl. Med. Biol., 1977, 86B, 85.
48. Asquith, R. S.; Otterburn, M. S. In "Protein Crosslinking. Nutritional and Medical Consequences"; (Friedman, M., Ed.), Plenum Press: New York, Adv. Exptl. Med. Biol., 1977, 86B, 93.
49. Asquith, R. S.; Skinner, J. D. Textilveredlung, 1970, 5, 406.
50. Bohak, Z. J. Biol. Chem., 1964, 239, 2878.
51. Finley, J. W.; Friedman, M. In "Protein Crosslinking. Nutritional and Medical Consequences"; (Friedman, M., Ed.), Plenum Press: New York, Adv. Exptl. Med. Biol., 1977, 86B, 123.
52. Horn, M. J.; Jones, D. B.; Ringel, S. J. J. Biol. Chem., 1941, 138, 141.
53. Spiro, R. G. Methods Enzymol., 1972, 28, 3.
54. Walsh, R. G. Ph.D. Thesis, Univ. Calif., Davis, 1978.
55. Phillip, M.; Polgar, L.; Bender, M. L. Methods Enzymol., 1970, 19, 215.
56. Ebert, Ch.; Ebert, G.; Rossmeissl, G. In "Protein Crosslinking. Nutritional and Medical Consequences"; (Friedman, M., Ed.), Plenum Press: New York, Adv. Exptl. Med. Biol.,

1977, 86B, 205.
57. Patchornik, A.; Sokolovsky, M. J. Am. Chem. Soc., 1964, 86, 1206.
58. Sokolovsky, M.; Sadeh, T.; Patchornik, A. J. Am. Chem. Soc., 1964, 86, 1212.
59. Patchornik, A.; Sokolovsky, M. Peptides, Proc. European Symp., 5th, Oxford, 1962, 253.
60. Edelhoch, H. Biochemistry, 1967, 6, 1948.

RECEIVED October 18, 1979.

Amino Acid Racemization in Alkali-Treated Food Proteins—Chemistry, Toxicology, and Nutritional Consequences

PATRICIA M. MASTERS

Scripps Institution of Oceanography, University of California, La Jolla, CA 92093

MENDEL FRIEDMAN

Western Regional Research Center, Science and Education Administration, U.S. Department of Agriculture, Berkeley, CA 94710

Food proteins are commonly treated with heat and occasionally with alkali during commercial and home processing. These treatments are intended to alter flavor and texture, destroy microorganisms, enzymes, toxins, or proteolytic enzyme inhibitors, and prepare protein concentrates. Undesirable changes also occur in the amino acid composition of proteins under such processing conditions. Amino acid crosslinking, degradation, amino acid–sugar complex formation, and racemization have been reported. Treated proteins have reduced digestibility, can produce symptoms of protein deficiency when fed to laboratory animals, and have been implicated in the etiology of rat kidney lesions.

Heat and alkaline treatments have been known since the early part of the century to racemize amino acid residues in proteins (1,2,). Dakin and Dudley (3) also studied digestibility of casein in vitro and in vivo after hydroxide treatment. Heating casein with 0.5 N NaOH at 37° for about 30 days completely prevented enzymatic hydrolysis and intestinal absorption when the treated casein was fed to a dog. The kinetics of base-catalyzed racemization of proteins was investigated by Levene and Bass (4-6). In these early studies, the extent of racemization was measured by changes in optical rotation.

More recently, enzymatic, microbiological, and chromatographic techniques have been used to determine extent of racemization. D-Lysine has been found in a sunflower protein

0-8412-0543-4/80/47-123-165$07.25/0

isolate treated in solutions of NaOH more concentrated than
0.2 N at 80° (7). Using microbiological assays, Tannenbaum
et al. (8) found that methionine was nearly completely racemized in fish protein concentrate heated 20 minutes at 95°
in 0.2 N NaOH. Heat treatment alone can racemize (or
epimerize) amino acid residues in proteins. Formation of
alloisoleucine was reported in bovine plasma albumin roasted
at 145° for 27 hours (9). Gas chromatography has been used
by Hayase et al. (10) to measure the racemization of eight
amino acids in roasted casein, lysozyme, and two poly-L-amino
acids.

Racemization is thought to proceed by abstraction of the
α-proton from an amino acid or amino acid residue in a peptide
or protein to give a negatively charged planar carbanion (11;
Figure 1). A proton can then be added back to either side of
this optically inactive intermediate, thus regenerating the
L-form or producing the D-enantiomer. The reaction can be
written as

$$\text{L-amino acid} \underset{k'}{\overset{k}{\rightleftharpoons}} \text{D-amino acid} \qquad (1)$$

where k and k' are the first order rate constants for inter-
conversion of the L- and D-enantiomers. Only L-amino acids
are initially present in most proteins due to the stereospecificity of biosynthesis.

In this paper we describe some of the factors which
influence racemization of amino acid residues in food proteins
and discuss toxicological and nutritional consequences of feeding alkali-treated food proteins.

Experimental

Alkali Treatment. The following is a typical procedure.
A 1% solution of each protein in 0.1 N NaOH (pH∿12.5) was
placed in a glass-stoppered Erlenmeyer flask and incubated at
the appropriate temperature in a water bath. The final pH
did not differ significantly from the initial value. After
three hours, the sample was dialyzed against 0.01 N acetic
acid for approximately two to three days and lyophilized.
Control protein samples were dialyzed and lyophilized similarly. The pH was measured with a Corning pH meter before
and after treatment.

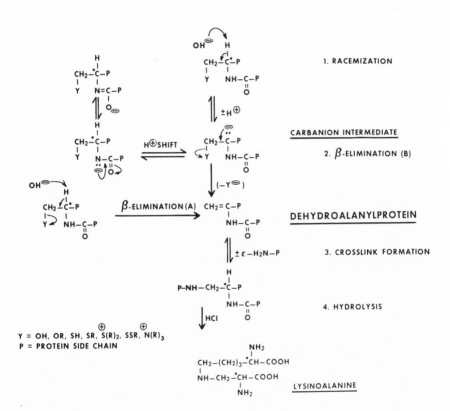

Figure 1.   Postulated mechanism of racemization and lysinoalanine formation via a common carbanion intermediate.   Note that two β-elimination pathways are possible: (a) a concerted, one-step process (A) forming the dehydroprotein directly; and (b) a two-step process (B) via a carbanion intermediate.   The carbanion, which has lost the original asymmetry, can recombine with a proton to regenerate the original amino acid residue which is now racemic.   Proton transfer may take place from the environment of the carbanion or from adjacent NH groups, as illustrated.   Protein anions and carbanions can also participate in nucleophilic addition and displacement reactions (24, 82, 83).

Enantiomeric Analyses (12). Approximately 30 mg of protein was hydrolyzed in 12 ml double–distilled 6 N HCl for 24 hours at 96°C. The HCl solution was then brought to dryness at 50°C under partial vacuum in a rotatory evaporator. The residue was redissolved in double–distilled water and desalted on Dowex 50W–X8 (100–200 mesh) resin that had been cleaned with NaOH and protonated with double–distilled HCl (13). Amino acids were eluted from the column with 1.5 M NH₄OH prepared by bubbling NH₃ through double distilled water. The effluent was again evaporated to dryness. Aspartic acid was separated by chromatography on a calibrated column of BioRad AG1–X8, 100–200 mesh, anion exchange resin. The resin was regenerated with 4 column volumes of 1 M sodium Half of the sample was applied to the column; elution was carried out with 1 M acetic acid. The aspartic acid fraction was evaporated in a rotatory evaporator. The L–leucyl–DL–aspartic acid dipeptides were synthesized by the procedure of Manning and Moore (14). The D/L aspartic acid ratio was determined with a Beckman Model 118 Automatic Amino Acid Analyzer.

D/L enantiomeric ratios for alanine, valine, glutamic acid, leucine, proline, and phenylalanine in the remaining half of the desalted amino acid fraction were obtained by gas chromatography (15). The N–trifluoroacetyl–L–prolyl–DL–amino acid esters were synthesized, then separated on a Hewlett–Packard Model 5711A Gas Chromatograph with flame ionization detector and a 12 foot column of Chromasorb W–AW–DMCS solid support coated with 8% SP 2250.

Amino Acid Analyses. A weighed sample (about 5 mg) of protein was hydrolyzed in 15 cc of 6 N HCl in a commercial hydrolysis tube. The tube was evacuated, placed in an acetone–dry ice bath, evacuated and refilled with nitrogen twice before being placed in an oven at 110°C for 24 hours. The cooled hydrolysate was filtered through a sintered disc funnel, evaporated to dryness at 40°C with the aid of an aspirator, and the residue was twice suspended in water and evaporated to dryness. Amino acid analysis of an aliquot of the soluble hydrolysate was carried out on a Durrum Amino Acid Analyzer, Model D-500 under the following conditions: single column ion–exchange chromatography method; Resin, Durrum DC-4A; buffer pH, 3.25, 4.25, 7.90; photometer, 440 nm, 590 nm; column, 1.75 mm X 48 cm; analysis time, 105 min. Norleucine was used as an internal standard.

In this system, lysinoalanine (LAL) is eluted just before histidine (16). The color constant of LAL was determined with an authentic sample purchased from Miles Laboratories.

Acylated proteins were prepared as described previously (16).

Results and Discussion

Racemization of Protein-Bound Amino Acids. D/L enantiomeric ratios for seven amino acid residues are given in
Table I. Extensive racemization of aspartic acid, phenylalanine, glutamic acid, and alanine occurred when the four
proteins were treated with hydroxide. Valine, leucine, and
proline were much less racemized.

The percentage of D-enantiomers relative to the total
amount of the amino acid residue can be calculated by the
relation (D/D+L) x 100. D-Aspartic acid accounts for 30% of
that residue (which is thus 60% racemized) in treated casein,
Promine-D, and wheat gluten. In these three proteins, 22-30%
of the phenylalanine (an essential amino acid) is the
D-enantiomer, and in wheat gluten, 26% of glutamic acid has
been converted to the D-form.

The small amounts of D-enantiomers in the controls may
be attributed to: (a) racemization occurring in the commercial preparation of the proteins, (b) acid-catalyzed
racemization during the hydrolysis step of our analysis, or
both factors.

Racemization rates (Table I) clearly differ among these
seven amino acids. To compare results from the four proteins,
rate constants were calculated from these data. For casein,
D/L ratios were measured at 0, 1, 3, 8 and 24 hours. These
results are plotted in Figure 2. The curves have two regions
of different racemization rates. Rapid initial rates observable up to about 3 hours are followed by slower rates up to
24 hours. The amino acids apparently have not reached equilibrium by 24 hours of incubation. Theoretically for amino
acids having one asymmetric center, the equilibrium D/L ratio
is 1.0. This value has been observed in fossil bone protein
(see 13) and in dry roasted proteins (10), but not in calcareous marine sediments (17) nor in fossil mollusc shell (18).
The linear first-order equation for the reversible reaction
can be used in the analysis of these results if the two
regions of the curves are treated separately. Initial rate
constants were estimated from the 0- and 3-hour points
using the equation

$$\left[\frac{1 + (D/L)}{1 - K' \, (D/L)} \right] = (1 + K') \cdot k \cdot t \qquad (2)$$

where $K' = 1/K_{eq}$ and K_{eq} is the D/L ratio at equilibrium

TABLE I. Enantiomeric Ratios In Hydroxide-Treated And Untreated Proteins

Treated Proteins[a]	Time (hrs)	D/L Asp	D/L Ala	D/L Val	D/L Leu	D/L Pro	D/L Glu	D/L Phe
Casein	1	0.279	0.044	0.028	0.053	0.031	0.111	0.191
	3	0.432	0.154	0.065	0.075	0.056	0.210	0.286
	8	0.489	0.241	0.079	0.157	0.040	0.350	0.439
	24	0.733	0.42	0.20	0.19	0.04	0.48	0.57
Promine-D	3	0.431	0.187	0.071	0.087	0.061	0.232	0.331
Wheat Gluten	3	0.409	0.156	0.040	0.059	0.033	0.349	0.304
Lactalbumin	3	0.293	0.101	0.050	0.061	0.037	0.139	0.198
Controls								
Casein	0	0.022	0.023	0.021	0.023	0.033	0.018	0.029
Promine-D	0	0.023	0.021	0.027	0.034	0.033	0.018	0.023
Wheat Gluten	0	0.034	0.020	0.021	0.018	0.033	0.021	0.024
Lactalbumin	0	0.032	0.022	0.030	0.028	0.032	0.030	0.023

(see 17 and Appendix for derivation). We are assuming provisionally that K_{eq} = 1, but further work is in progress to elucidate this point.

The rates within each protein were then standardized relative to that of leucine. The order of relative racemization rates is presented in Table II. Relative rates are very similar among the various proteins except for aspartic acid and glutamic acid in wheat gluten. This situation is discussed below. (The relative rate constants estimated for the second region of the casein curves in Figure 2, using the 3-hour and 24-hour points, is k(asp) : k(phe) : k(glu):k(ala) : k(leu) = 4.0 : 3.0 : 2.5 : 2.5 : 1.0.)

Interpreting the kinetics of base–catalyzed racemization in these proteins is complicated by the simultaneous hydrolysis of the original proteins during the experiment. By 24 hours, the amount of recoverable dialysate is roughly 25–50% by weight of the starting material. The dialysis procedure eliminates lower molecular weight species as the incubation proceeds. Varying rates of isoleucine racemization (actually epimerization) have been attributed to differences in protein composition (18). Consequently, the two regions of the curves in Figure 2 may represent rates in two or more distinct populations of molecules resulting from hydrolysis and dialysis.

Other interpretations of the data in Figure 2 also may be considered. The rapid initial rates could reflect a more ready formation of the carbanion during early stages of the denaturation process. Neighboring groups have been shown to influence racemization rates during acid hydrolysis (14). Conceivably, as native conformation is altered by exposure to high temperature and hydroxide, different sets of residues will be brought into proximity with any particular amino acid. Alternatively, some residues may simply be more susceptible to racemization, possibly because of the primary structure. These labile residues then form carbanion intermediates rapidly in the strongly dissociative solvent environment.

The results in Table II can be compared with information on free amino acids. Bada (19) has shown that for free amino acids at neutral pH, k(ileu): k(val): k(ala): k(phe): k(asp) = 1.0 : 0.8 : 2.4 : 4.4 : 8.6. He has pointed out that these rates agree with the order predicted from σ^* values of the respective R–groups (20). The R–groups that have the greater electron–withdrawing or resonance–stabilizing characteristics will induce faster racemization. Our results are compatible with the free amino acid data.

It is interesting that even at pH ∿ 12.5, the relative order of initial racemization rates in bound amino acids appear very similar to those of free amino acids at neutral pH. Since both the NH_2 and COOH groups are involved in peptide bonds, their ionic forms are no longer relevant in the reaction

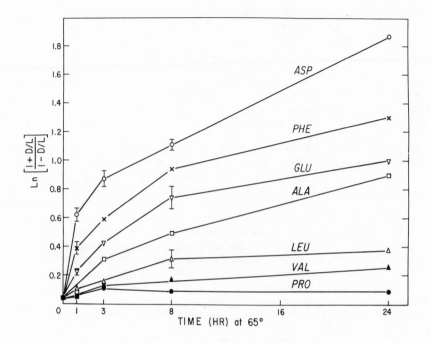

Figure 2. Time course of amino acid racemization reactions of casein in 0.1N NaOH at 65°C.

TABLE II. Order Of Initial Racemization Rates Relative to

Leucine.   Proteins Treated With 0.1 N̲ NaOH AT 65°C

	k_{leu} :	k_{val} :	k_{pro} :	k_{ala} :	k_{glu} :	k_{phe} :	k_{asp}
Casein	1.0	0.9	0.5	2.6	3.7	4.9	7.8
Promine-D	1.0	0.8	0.5	3.1	4.1	5.9	8.1
Wheat Gluten	1.0	0.4	0	3.3	8.2	6.9	9.6
Lactalbumin	1.0	0.6	0.2	2.5	3.4	5.4	8.2

mechanism. The R-substituents remain as the primary influence on relative racemization rates on proteins. At neutral and basic pH, the predominant ionic form of the β-carboxyl group of aspartic acid (and of the γ-carboxyl of glutamic acid) is COO^-. Although more electronegative than the undissociated carboxyl groups, these ionized carboxyl groups still have greater electron withdrawing capacity than do the alkyl groups of alanine, leucine and valine. Phenylalanine, with an R-group σ^* close to that of the β-carboxylate, may have a slower racemization rate relative to aspartic acid due to steric limitations on the formation of the planar carbanion intermediate. The protein-bound imino acid proline would be even more limited sterically.

The correlation between racemization rates in free amino acids and the σ^* values also supports the carbanion-intermediate mechanism of racemization (17). The R-group can act to stabilize the negative charge on the α-carbon so that the carbanion intermediate is more stable. Since the σ^* values also agree with the racemization rates observed in the present study, the same mechanism probably operates with protein-bound amino acids. It is noteworthy, however, that the racemization rate of free aspartic acid is $\sim 10^{-5}$ relative to those reported here for this amino acid residue in proteins (17-19). (For relevant discussions on the influence of R groups on reactivities of amino acids, peptides, and proteins, see references 21-26).

Where a relative rate differs from the observed pattern, as is the case of glutamic acid in wheat gluten, the apparent rate enhancement may be the result of the very high proportion of glutamine in the protein (27). The δ-amide group should increase the inductive character of the R-substituent so that glutamine should racemize faster than glutamic acid. Because glutamine (and asparagine) are probably deamidated to some extent during the alkali treatment and completely during acid hydrolysis of the proteins, our D-glutamic acid values actually represent the sum of both D-amino acid enantiomers. This circumstance may explain the relative rate difference for glutamic acid in wheat gluten if deamidation during hydroxide treatment is slower than racemization under our experimental conditions.

Deamidation of glutamine and asparagine is sequence-dependent with half-times ranging from 18 to 507 days for asparaginyl peptides and 96 days for glutaminyl residues at 37°, pH 7.4 (28). The pH dependence of the deamidation rates was studied in phosphate buffer (29). Extrapolation of the curve for one glutaminyl peptide to pH 12 (approximating the conditions used in our study) results in a $k_{deam} \simeq 2x$ that of the expected k_{rac} of glutamic acid. Since the majority of the 24 sequences studied (28) deamidate more slowly than this peptide, it seems probable that the majority of glutamines will remain intact during the 3 hour time course of our treatments.

Although relative racemization rates of the same amino acids in different proteins are usually similar, the overall lability of these four proteins to racemization by hydroxide differs considerably. In Table III, the proteins are ranked by the extent of racemization after 3 hr at 65°C of each of the four most racemized amino acids. Promine-D is the most highly racemized protein for three of these amino acids. Lactalbumin has the lowest D/L ratios for all. These findings imply that there are protein-specific rates of racemization underlying the general uniformity of relative rates discussed above. Similar observations have been reported for diagenetic racemization rates in different fossil proteins (18, 30, 31). Variability among these food proteins in response to alkali treatment demonstrates that moderate processing conditions for one protein (e.g. lactalbumin) constitute more severe conditions for another (e.g. Promine-D).

The pH dependence of the racemization rate of aspartic acid in casein was also investigated. The results are plotted in Figure 3. Racemization rates are estimated from the log conversion of the D/L ratios. The pH of the NaOH buffer at 65° was calculated from the temperature variation of the pK_w of water (32). The pH values of the borate buffers at 65° were calculated using the temperature data of Bates (33). The solid line represents rates which are first order with respect to hydroxide concentration. This line is a reasonable fit to the data points above pH 10. Further experiments are in progress with other proteins in order to identify the lowest OH^- concentrations that induce first-order racemization kinetics. If the critical base concentrations for racemization correspond with the different responses of the four proteins (see Tables I and III), one may expect k(rac) for Promine-D to become first order at lower OH^- concentration than casein, while that of wheat gluten should be about the same as for casein. Lactalbumin may have the highest OH^- concentration tolerance.

Data from additional studies gave a plot of log k_{asp} vs. 1/T in the temperature range 25 to 75°C that was linear with an activation energy of 21.9 kcal/mole (Figures 4 and 5). This k value is similar to that obtained by Darge and Thiemann (34) for racemization of both free and bound amino acids in alkaline solutions. We are presently attempting to compare activation energies for racemization of other amino acid residues in proteins with known values for free amino acids.

TABLE III.   Relative Order Of The Proteins With Respect To

Extent Of Racemization Of Each Amino Acid[a]

Asp	Promine-D	=	Casein	>	Wheat Gluten	>	Lactalbumin
Ala	Promine-D	>	Casein	=	Wheat Gluten	>	Lactalbumin
Glu	Wheat Gluten	>	Promine-D	>	Casein	>	Lactalbumin
Phe	Promine-D	>	Wheat Gluten	>	Casein	>	Lactalbumin

[a]After 3 hr at 65°C.

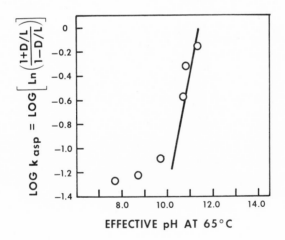

Figure 3. The pH dependence of aspartic acid racemixation in casein in the pH range 8 to 14. The D/L ratios are plotted as a function of the actual pH values calculated by methods explained in the text. The expression $\log k_{asp} \simeq \log[\ln(1 + D/L/1 - D/L)]$ is derived from eq 2. Under our experimental conditions, eq 2 can be reduced to this one term when solving for k (12). The line is a first-order kinetics plot superimposed on the calculated data points.

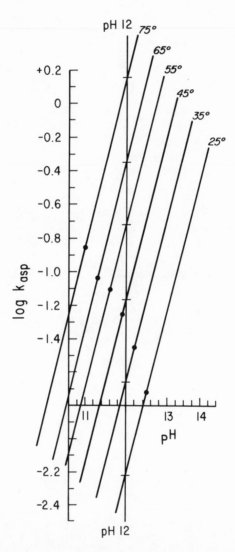

Figure 4. A nomogram for pH corrections for temperature kinetics of racemiza-tion in commercial casein (see text). Since effective pH is dependent on tempera-ture, all k_{asp} values are adjusted to pH 12 using first-order extrapolation lines, as shown. These adjusted log k_{asp} values are then plotted in Figure 5.

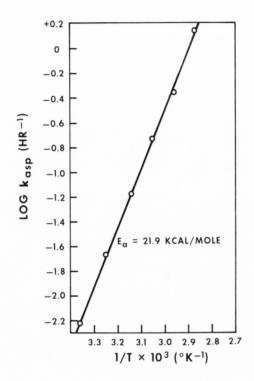

Figure 5. Arrhenius plot of Figure 4 values. Racemization rates for casein were determined at 10°C intervals in the temperature range 25°–75°C.

Discriminating Between the Effects of Racemization and
Lysinoalanine Formation. A mechanism of racemization and
lysinoalanine formation is illustrated in Figure 1.
Racemization occurs by abstraction of a proton by hydroxide
ion (or any general base) from an optically-active carbon atom
to form a carbanion, that has lost the original asymmetry.
The pair of unshared electrons on the carbanion can undergo
two reactions: (a) it can recombine with a proton from the
solvent to regenerate either the original amino acid side
chain or its optical antipode, so that it is racemized; (b)
it can undergo the indicated β-elimination reaction to form
a dehydroalanine derivative, which can then combine with an
ε-amino group of a lysine side chain to form a lysinoalanine
crosslink.

Since lysinoalanine and at least one D-amino acid are
toxic to some animals (35), we wished to distinguish their
effects in alkali-treated proteins. Such discrimination is
possible, in principle, since we have found that acylating
the ε-amino group of lysine proteins seems to prevent
lysinoalanine formation. Since lysinoalanine formation from
lysine requires participation of the ε-amino group of lysine
side chains, acylation of the amino group with acetic anhydride is expected to prevent lysinoalanine formation under
alkaline conditions if the protective effect survives the
treatment. This is indeed the case (16).

Although acetylation appears to minimize or prevent
lysinoalanine formation, our findings indicate that acylation
does not significantly change the extent of racemization after
3 hr at 65°C (Table IV). These results show that it is possible in principle to discriminate between the alkali-induced
effects of racemization and lysinoalanine formation. Additional studies in progress are designed to further delineate
this principle.

Cytotoxic and Therapeutic Consequences. Alkali-treated soy
proteins, when fed to rats, induce changes in kidney cells
characterized by enlargement of the nucleus and cytoplasm,
increase in nucleoprotein, and disturbances in DNA synthesis
and mitosis. The lesion, first described by Newberne and Young
(36) and attributed to lysinoalanine by Woodard (37,49) and
now designated nephrocytomegaly, affects the epithelial cells
of the straight portion (pars recta) of the proximal renal
tubules. Renal cytomegaly of the pars recta is also induced by
feeding rats synthetic lysinoalanine (38-40). Since this
unusual, crosslinked amino acid is formed in proteins during
alkaline treatment, the nephrotoxic action of the treated proteins was ascribed to the presence of lysinoalanine (LAL).

Enlarged nuclei tend to have more than the diploid complement of DNA, unusual chromatin patterns, and proteinaceous
inclusions (37). Increases in total nonchromosomal protein

parallel the increases in nuclear volume. These observations indicate disruption of normal regulatory functions in the <u>pars recta</u> cells.

The spectrum of cytological changes described above also appears in response to the renal carcinogens: aflatoxin (<u>41</u>). lead (<u>42</u>), 4'-fluoro-4-amino diphenyl (<u>43</u>), N(4-fluoro-4-biphenyl) acetamide (<u>44</u>), and dimethylnitrosamine (<u>45</u>). Nephrocytomegaly has been used by investigators as an indicator of pre-neoplastic conditions (<u>46,47</u>). Feeding experiments with rats have not been followed over long enough periods, however, to test the tumorigenicity of alkali-treated soy proteins (<u>48</u>).

Apparent divergent findings by several laboratories suggest that factor(s) other than LAL content may modify the biological response to alkali-treated proteins. Diets containing synthetic LAL or alkali-treated proteins with high LAL content did not produce renal cytomegaly in mice, hamsters, rabbits, quail, dogs or monkeys (<u>39</u>). With rats, de Groot and coworkers (<u>39</u>, <u>49</u>, <u>50</u>) initially failed to observe nephrocytomegaly when feeding alkali-treated soy protein (although they did report the renal changes when feeding free LAL (<u>39</u>, <u>50</u>)). This contrasts with the findings of Woodard and others (<u>37</u>, <u>38</u>, <u>51</u>, <u>52</u>), Newberne and others (<u>41</u>, <u>53</u>), Struthers et al. (<u>54</u>), and Karayiannis (<u>55</u>) who have all seen the renal lesions in rats fed alkali-treated soy proteins.

Another difficulty in formulating a simple relationship between LAL and nephrocytomegaly is that proteins of equivalent LAL content produce different biological responses. Some laboratories (<u>38</u>, <u>55</u>) reported severe nephrocytomegaly from alkali-treated soy protein while a similar protein, having the same LAL content, does not produce lesions (<u>56</u>). Karayiannis (<u>55</u>) has found that alkali-treated soy protein (supplying 1400-2600 ppm LAL) resulted in nephrocytomegaly whereas 2500 ppm LAL derived from alkali-treated lactalbumin did not induce the lesions.

The divergent observations about relative potencies of various alkali-treated proteins in inducing kidney lesions could arise from dietary factors since adding high-quality, untreated proteins such as casein and lactalbumin to diets containing alkali-treated proteins appears to prevent the lesion (<u>57</u>, <u>58</u>). One possible explanation for this effect is that amino acids (<u>e.g.</u> lysine, methionine) derived from added proteins may prevent the binding of lysinoalanine to metalloproteins such as metallothioneins present in the kidney 59-61). The observed long residence time of lysinoalanine in the kidney (<u>62</u>) may be related to a possible chelating action of lysinoalanine, which has three amino and two carboxyl groups that could participate in binding metal ions of metallo proteins.

The conflicting reports from various laboratories might also be explained by amino acid racemization during alkaline processing of the test proteins. Our studies show that four different proteins subjected to the same alkaline treatment exhibited varying degrees of racemization. One result of the presence of D-amino acids would be to decrease enzymatic digestion of the proteins, thus restricting the amount of free LAL released. Reduced digestibility has been observed with proteins subjected to severe alkaline treatment (49, 50). In order of cytotoxic effect, the most pronounced response is with free LAL, then low molecular weight LAL-containing peptides, then LAL-containing proteins (56,63, 64). Thus, different proteins having the same LAL content will be expected to release differing quantities of free LAL depending upon their extents of racemization. Even closely related proteins such as α- and γ-crystallins from calf lens racemize at different rates (P. M. Masters, unpublished results). Some of the discrepancies thus may be attributed to the use of different fractions of soy protein in the experimental diets. Note also that the alanine part of lysinoalanine is a potential precursor for D,L-serine and that D-serine and one of the lysinoalanine diasteroisomers offer similar configurations to potential receptor sites (35, 58, 59).

We have also shown that small differences in the conditions of alkaline treatment can produce fairly large differences in the extent of racemization in casein. Temperature, of course, and length of treatment are critical. However protein concentration does not appear to influence significantly the extent of racemization (Table V). Therefore, treatment conditions may generate comparable contents of LAL but varying D-amino acid contents.

D-Amino acids may have other effects in nephrocytomegaly in addition to influencing release of LAL. Soy protein, the most labile to racemization of the four proteins studied, is more cytotoxic than lactalbumin (48, 58), which is the least racemized. If, as postulated, D-amino acids were seriously inhibiting release of LAL, then lactalbumin would be expected to yield more free LAL than soy protein.

Karayiannis (55) has reported that diets supplying 5000 ppm LAL in alkali-treated lactalbumin produced only mild cytomegaly, whereas 1400-2600 ppm LAL in treated soy protein produced extensive cytotoxicity. Since the lactalbumin was treated for 80 minutes at 60°C (58), little racemization (relative to the soy protein at 60°C for 8 hrs) would have occurred. The cytotoxicity of soy protein may be due to something besides, or in addition to, LAL content. It is possible that D-amino acid(s) may act synergistically with LAL in the expression of nephrocytomegaly. It has been known for sometime that D-serine can induce renal lesions when fed to rats

TABLE IV. Aspartic Acid Racemization In Modified
Casein: 1% Protein, 0.1 $\underline{N}$ NaOH, 65°, 3 Hr

Sample	D/L
Acetylated	0.336
Citraconylated	0.302
Glutarylated	0.343
Unmodified	0.348

TABLE V. Effect Of Protein Concentration On
Aspartic Acid Racemization In Casein:
0.1 $\underline{N}$ NaOH, 75°, 3 Hr

Concentration	Final pH	D/L Asp
0.5%	12.5	0.466
2.0%	12.4	0.596
5.0%	12.4	0.526
9.6%	12.3	0.546

(35). Another possibility is that some unknown factor may be responsible for the lesions.

If the protein-induced renal cytomegaly is precancerous, it is a cause for concern since alkali-processed protein is used extensively in commercial food preparations (56,63,64). Although lysinoalanine concentrations in foods are usually lower than amounts needed to induce nephrocytomegaly in rats, health hazard may exist since human tolerances for the "unnatural" amino acids generated during commercial processing are not known. If the protein-induced lesion is not precancerous, it is still important to understand its etiology.

Maillard products such as lysine-sugar complexes are generated by heating proteins in the presence of carbohydrates. When casein heated with glucose under mild conditions (to maximize formation of ε-fructosyllysine without inducing LAL crosslinks) was fed to rats, histopathological changes resembling nephrocytomegaly were observed (65). According to preliminary evidence, heating free aspartic acid in the presence of glucose increases the racemization rate two- to threefold (66). Therefore, significant amounts of racemization may be induced under these conditions. Thus, the cytotoxic response observed by Erbersdobler et al. (65) may be due to D-enantiomers as well as to Maillard products. A more exact understanding of the relation of feed or food quality to processing conditions depends, to an important degree, on more information about the occurrence and conditions of formation of "unnatural" amino acid derivatives.

Nutritional Implications. The nutritive quality of any protein depends on three factors: amino acid composition, digestibility, and utilization of the released amino acids. Racemization brought about by processing can impair the nutritive value of proteins by (a) generating non-metabolizable forms of amino acids (D-enantiomers), (b) creating peptide bonds inaccessible to proteolytic enzymes, and (c) toxic action (or interaction) of specific D-enantiomers. Little is known concerning the health consequences of human consumption of racemized proteins. No study has specifically evaluated amino acid losses due to racemization within food proteins.

A major consideration is whether humans can utilize the D-enantiomers of essential amino acids. Berg (67) has reviewed human and animal utilization of free D-amino acids. L-Amino acids are invariably taken up faster than the D-enantiomers in the intestine (68, 69) and kidney (70). Once absorbed, D-amino acids can be utilized by two pathways: (a) racemases (or epimerases) may convert D-isomers to racemic mixtures; or (b) D-amino acid oxidases may catalyze oxidative deamination to α-keto acids, which can be specifically reaminated to the natural L-forms (71). Only the latter activity has been demonstrated in mammals. This enzyme system should permit human utilization of D-amino acids for growth and maintenance. Of the D-isomers of the eight essential amino acids, however, only D-phenylalanine and D-methionine were found to maintain human nitrogen equilibrium in early studies (72). More recently, Kies et al. (73) and Zezulka and Calloway (74) presented evidence that D-methionine is, in fact, poorly utilized by humans. When mixtures of D-amino acids are fed to rats, the oxidase system can be overloaded so that the D-enantiomers of essential amino acids cannot be transaminated in sufficient quantity to support growth (75).

Physiological accessibility of the essential amino acids also influences quality. Alkali-treated proteins have been shown to have reduced digestibility in vitro (7,49,76) and in vivo (49, 62) and to produce symptoms of dietary inadequacy such as poor growth, hair loss, and diarrhea (34, 54, 75). These studies focused attention on the crosslinked amino acid derivative, lysinoalanine (LAL), and its possible toxicity. Although the treatments used almost certainly caused significant racemization, the possible effect of D-amino acids on the results was not evaluated.

Since D-amino acids are poorly utilized, diets containing sufficient quantities of D-enantiomers will result in elevated levels of plasma and urinary amino acids. Urinary excretion of D-methionine by infants fed a formula supplemented with DL-methionine has led to misdiagnosis of inborn errors of metabolism (77). D-Amino acids derived from processed food proteins may confuse medical diagnoses. Determining D-amino acid contents of common foods would estimate the significance of this problem. Some preliminary results are shown in Table VI.

Finally, since our results show that aspartic acid and phenylalanine racemize at fast rates, recent reports suggest new dimensions in research on racemization of food proteins. First, an analgesic effect has been attributed to D-phenylalanine (78). The described therapeutic effect of D-phenylalanine is presumably due to its ability to inhibit an enzyme responsible for destroying the natural opiate-like enkephalins in the brain. Consequently, measurement of the extent of racemization and enzymatic release of D-phenylalanine and other amino acids in food proteins which may be taken up by brain tissue (79, 80) merits further study. A related question is whether the sweetening agent Aspartame (L-aspartyl-L-phenylalanine methyl ester, 81) undergoes racemization during food processing and cooking, leading to a decrease in its sweetening power.

TABLE VI.   D-ASPARTIC ACID CONTENT IN

COMMERCIAL FOOD PRODUCTS

Product	D/L Asp	% D-Asp[*]
Coffee-Mate	0.208	17%
Plus Meat (textured soy protein)	0.095	9%
Fritos	0.164	14%
Isomil	0.108	10%
Breakfast Strips (simulated bacon)	0.143	13%

[*] Calculated as (D/D+L) x 100. This gives the relative percentage of aspartic acid as D- and L-enantiomers, not the absolute percentage in the food.

Summary

Alkali treatment of proteins catalyzes racemization of optically active amino acids. A study of the effect of alkali on commercial wheat gluten, soy protein, casein, and lactalbumin showed that racemization rates vary among proteins but that, within each protein studied, the relative order is similar. Factors which influence racemization include pH, temperature, time of exposure to alkali, and the inductive nature of amino acid side chains. Protein-bound D-amino acids formed during alkali and heat treatment of food proteins may adversely affect the nutritional quality and safety of processed foods. This may be the result of decreased amounts of essential amino acid L-enantiomers, decreased digestibility through peptide bonds not susceptible to normal peptidase cleavage, specific toxicity of certain D-isomers, and/or modification of the biological effects of lysinoalanine or other unnatural amino acids.

Acknowledgment

We thank S. Steinberg and J. R. Whitaker for helpful comments.

Appendix

First Order Kinetic Equations for Reversible Amino Acid Racemization

$$L \underset{k'}{\overset{k}{\rightleftharpoons}} D$$

$-dL/dt = kL - k'D$

if $D_{t=0} \ll L_{t=0}$, then $L_{t=0} = L + D$ and $L_{t=0} - L = D$

$-dL/dt = kL - k' (L_{t=0} - L) = kL - k'L_{t=0} + k'L$

$\qquad = (k + k')L - k'L_{t=0}$

$dL/dt = - (k + k')L + k'L_{t=0}$

$dL/dt + (k + k')L = k'L_{t=0}$

$dL/dt \, e^{(k + k')t} + (k + k')L \, e^{(k+k')t} = k'L_{t=0} \, e^{(k +k')t}$

$d/dt \, (Le^{(k + k')t}) = k'L_{t=0} \, e^{(k + k')t}$

$\int d \, (Le) \, (Le^{(k + k')t}) = \int (k'L_{t=0} \, e^{(k + k')t}) \cdot dt$

$Le^{(k +k')t} = [k/(k + k')]L_{t=0} \, e^{(k + k')t} + \text{constant}$

$L = [k'/(k + k')]L_{t=0} + \text{constant} \cdot e^{-(k + k')t}$

At $t = 0$, $L = L_{t=0}$ and $e^{-(k + k')t} = 1$

$L_{t=0} = [k'/(k + k')]L_{t=0} + \text{constant}$

$L_{t=0} - [k'/(k + k')]L_{t=0} = \text{constant}$

$L_{t=0}[(1 - k'/(k + k')] = \text{constant}$

$L = [k'/(k + k')]L_{t=0} + L_{t=0} \, k/(k + k') \, e^{-(k +k')t}$

if $L_{t=0} = L + D$

$L = [k'/(k + k')](L + D) + (L + D) \, k/(k + k') \, e^{-(k + k')t}$

$$L/(L + D) = k'/(k + k') + [k/(k + k')]e^{-(k - k')t}$$

$$\frac{L(k + k') - k'(L + D)}{(L + D)(k + k')} = [k/(k + k')]e^{-(k + k')t}$$

$$= \frac{Lk - Dk'}{(L + D)(k + k')}$$

$$\left[\frac{L - D(k'/k)}{L + D}\right] = e^{-(k + k')t}$$

$$= \left[\frac{L[(1 - (D/L)(k'/k)]}{L[(1 + (D/L)]}\right]$$

$$= \left[\frac{1 - (D/L)(k'/k)}{1 + (D/L)}\right]$$

$$\ln\left[\frac{1 - (D/L)(k'/k)}{1 + (D/L)}\right] = -(k + k')t$$

$$\ln\left[\frac{1 + (D/L)}{1 - (D/L)(k'/k)}\right] = (k + k')t$$

if $k'/k = K'$

$$\ln\left[\frac{1 + (D/L)}{1 - K'(D/L)}\right] = (1 + K') \cdot k \cdot t \qquad (eq. 2)$$

An alternate mathematical treatment of the first-order kinetics of racemization gives equation 3. However, equation 2 is operationally more efficient than equation 3 because we can measure D/L ratios with a 1-3% error compared to a 10-15% error when concentrations of D only are measured.

$$\ln (D_e/D_e - D_t) = (k + k')t = k_{(obs)}t \qquad (eq. 3)$$

where D_e = equilibrium value of D

and D_t = D at time t.

Literature Cited

1. Kossel, A., and Weiss, F. (1909). Über Einwirkung von Alkalien auf Proteinstoffe. Z. Physiol. Chem. 59, 492–498.

2. Dakin, H. D. (1912–1913). The racemization of proteins and their derivatives resulting from tautomeric change. J. Biol. Chem. 13, 357–362.

3. Dakin, H. D. and Dudley, H. W. (1913). The action of enzymes on racemized proteins and their fate in the animal body. J. Biol. Chem. 15, 271–276.

4. Levene, P. A. and Bass, L. W. (1927). Studies on racemization. V. The action of alkali on gelatin. J. Biol. Chem. 74, 715–725.

5. Levene, P. A. and Bass, L. W. (1928). Studies on racemization. VII. The action of alkali on casein. J. Biol. Chem. 78, 145–157.

6. Levene, P. A. and Bass, L. W. (1929). Studies on racemization. VIII. The action of alkali on proteins: racemization and hydrolysis. J. Biol. Chem. 82, 171–190.

7. Provansal, M.M.P., Cuq, J.-L.A. and Cheftel, J.-C. (1975). Chemical and nutritional modifications of sunflower proteins due to alkaline processing. Formation of amino acid crosslinks and isomerization of lysine residues. J. Agric. Food Chem. 23, 938–943.

8. Tannenbaum, S. R., Ahern, M., and Bates, R. P. (1970). Solubilization of fish protein concentrate. 1. An alkaline process. Food Tech., 24, 96–99.

9. Bjarnason, J. and Carpenter, K. J. (1970). Mechanisms of heat damage in proteins. Brit. J. Nutr. 24, 313–329.

10. Hayase, F., Kato, H., and Fujimaki, M. (1975). Racemization of amino acid residues in proteins and poly(L-amino acids) during roasting. J. Agric. Food Chem. 23, 491–494.

11. Neuberger, A (1948). Stereochemistry of amino acids. Adv. Protein Chem., 4, 298–383.

12. Masters, P. M. and Friedman, M. (1979). Racemization of amino acids in alkali-treated food proteins. J. Agric. Food Chem., 27, 507–511.

13. Bada, J. L., and Protsch, R. (1973). Racemization of aspartic acid and its use in dating fossil bones. Proc. Natl. Acad. Sci. U.S.A., 70, 1331-1334.

14. Manning, J. M., and Moore, S. (1968). Determination of D- and L-amino acids by ion-exchange chromatography as L-D and L-L dipeptides. J. Biol. Chem., 243, 5591-5597.

15. Hoopes, E. A., Peltzer, E. T., and Bada, J. L., (1978). Determination of amino acid enantiomeric ratios by gas liquid chromatography of the N-trifluoroacetyl-L-prolyl-peptide methyl esters. J. Chromatographic Sci., 16, 556-560.

16. Friedman, M. (1978). Inhibition of lysinoalanine synthesis by protein acylation. In "Nutritional Improvement of Food and Feed Proteins", M. Friedman, Ed., Plenum Press, New York, pp. 613-648.

17. Bada, J. L., and Schroeder, R. A. (1972). Racemization of isoleucine in calcareous marine sediments: kinetics and mechanism. Earth Planet. Sci. Lett., 15, 1-11.

18. Masters, P. M., and Bada, J. L. (1977). Racemization of isoleucine in fossil molluscs from indian middens and interglacial terraces in Southern California. Earth Planet. Sci. Lett., 37, 173-183.

19. Bada, J. L. (1972). Kinetics of racemization of amino acids as a function of pH. J. Amer. Chem. Soc., 94, 1371-1373.

20. Charton, M. (1964). Definition of "inductive" substituent constants. J. Org. Chem., 29, 1222-1227.

21. Friedman, M., and Wall, J. S. (1964). Application of a Hammett-Taft relation to kinetics of alkylation of amino acid and peptide model compounds with acrylonitrile. J. Amer. Chem. Soc. 86, 3735-3741.

22. Friedman, M. Cavins, J. F., and Wall, J. S. (1965). Relative nucleophilic reactivities of amino groups and mercaptide ions in addition reactions with α, β-unsaturated compounds. J. Amer. Chem. Soc. 87, 3572-3682.

23. Friedman, M., and Wall, J. S. (1966). Additive linear free energy relationships in reaction kinetics of amino groups with α, β-unsaturated compounds. J. Org. Chem. 31, 2888-

24. Krull, H., and Friedman, M. (1967). Anionic polymerization of methyl acrylate to protein functional groups. J. Polym. Sci. A-1, 5, 2535-2546.

25. Friedman, M., and Sigel, C. W. (1966). A kinetic study of the ninhydrin reaction. Biochemistry 5, 478-484.

26. Friedman, M. (1967). Solvent effects in reactions of amino groups in amino acids, peptides, and proteins with α, β-unsaturated compounds. J. Amer. Chem. Soc. 89, 4709-4713.

27. Kasarda, D. D., Bernardin, J. S., and Nimmo, C. C. (1976). Wheat proteins. In: "Adv. Cereal Science and Technol.", Y. Pomeranz, Ed., American Association of Cereal Chemists, St. Paul, Minnesota, pp. 158-236.

28. Robinson, A. B., Scotchler, J. W., and McKerrow, J. H. (1973). Rates of nonenzymatic deamidation of glutaminyl and asparaginyl residues in pentapeptides. J. Amer. Chem. Soc. 95, 8156-8159.

29. Scotchler, J. W., and Robinson, A. B. (1974). Deamidation of glutaminyl residues: dependence on pH, temperature, and ionic strength. Anal. Biochem., 59, 319-322.

30. Miller, G. H., and Hare, P. E. (1974). Use of amino acid reactions in some arctic marine fossils as stratigraphic and geochronological indicators. Carnegie Inst. Washington Yearbook, 74, 612-617.

31. King, K., Jr., and Neville, C., (1977). Isoleucine epimerization for dating marine sediments: importance of analyzing monospecific foraminiferal samples. Science, 193, 1333-1335.

32. Robinson, R. A., and Stokes, R. H., (1959). "Electrolyte Solutions", Butterworths, London, pp. 336-369.

33. Bates, R. G., (1964). "Determination of pH: Theory and Practice". John Wiley and Sons, New York, p. 76.

34. Darge, W. and Thiemann, W. (1971). Hydrolysis and racemization of oligopeptides studied by optical rotation measurement and ion exchange chromatography. In: "First European Biophysics Congress Proceedings", E. Broda, A. Locker, and H. Springer-Lederer, Eds., Baden, Vol. 1, 133-137.

35. Wachstein, M. (1947). Nephrotoxic action of D,L-serine in the rat. Arch. Pathol., 43, 503-514.

36. Newberne, P. M. and Young, V. R. (1966). Effects of diets marginal in methionine and choline with and without vitamin B_{12} on rat liver and kidney. J. Nutr. 89, 69-79.

37. Woodard, J. C. and Alvarez, M. R. (1967). Renal lesions in rats fed diets containing alpha protein. Arch. Path. 84, 153-162.

38. Woodard, J. C. and Short, D. D. (1973). Toxicity of alkali-treated soy protein in rats. J. Nutr. 103, 569-574.

39. De Groot, A. P., Slump, P., Feron, V. J. and Van Beek, L. (1976). Effects of alkali-treated proteins: feeding studies with free and protein-bound lysinoalanine in rats and other animals. J. Nutr. 106, 1527-1538.

40. Woodard, J. C., Short, D. D., Alvarez, M. R. and Reyniers, J. P. (1975). Biological effects of N^{ε}-(DL-2-amino-2-carboxyethyl)-L-lysine, lysinoalanine. In: "Protein Nutritional Quality of Foods and Feeds", M. Friedman, Ed., Part 2, Marcel Dekker, Inc., N.Y. pp. 595-618.

41. Newberne, P. M., Rogers, A. E. and Wogan, G. N. (1968). Hepatorenal lesions in rats fed a low lipotrope diet and exposed to aflatoxin. J. Nutr. 94, 331-343.

42. Zollinger, H.-U. (1953). Durch chronische Bleivergiftung erzengte Nierenadenome und -carcinome bei Ratten und ihre Beziehungen zu den entsprechenden Neubildungen des Menschen. Virchows Arch. Path. Anat. 323, 694-710.

43. Mathews, J. J. and Walpole, A. L. (1958). Tumors of the liver and kidney induced in Wistar rats with 4'-fluoro-4-amino diphenyl. Brit. J. Cancer 12, 234-241.

44. Heatfield, B. M., Hinton, D. E. and Trump, B. F. (1976). Adenocarcinoma of the kidney. II. Enzyme histochemistry of renal adenocarcinomas induced in rats by N-(4-fluoro-4-biphenyl) acetamide. J. Nat. Cancer Inst., 57, 795.

45. Zak, F. G., Holzer, H. J., Singer, E. J. and Pepper, H. (1960). Renal and pulmonary tumors in rats fed dimethyl-nitrosamine. Cancer Res. 20, 96-99.

46. Schoental, R. and Magee, P. N. (1959). Further observations on the subacute and chronic liver changes in rats after a single dose of various pyrrolizidine (Senecio) alkaloids. J. Path. Bact. 78, 471-482.

47. Sugihara, R. and Sugihara, S. (1976). Electron microscopic observations on the morphogenesis of renal cell carcinoma induced in rat kidney by dimethylnitrosamine and N-(3,5-dichlorophenyl) succinimide. Cancer Res. 36, 533-550.

48. Gould, D. H. and MacGregor, J. T. (1977). Biological effects of alkali-treated protein and lysinoalanine: an overview. In: "Protein Crosslinking: Nutritional and Medical Consequences," M. Friedman, Ed., Part B, Plenum Press, N.Y., pp. 29-48.

49. De Groot, A. P. and Slump, P. (1969). Effects of severe alkali treatment of proteins on amino acid composition and nutritive value. J. Nutr. 98, 45-56.

50. De Groot, A. P., Slump, P., Van Beek, L. and Feron, V. J. (1976). Severe alkali treatment of proteins. In: "Evaluation of Proteins for Humans," C. E. Bodwell, Ed., Avi Publishing Co., Westport, CT, pp. 270-283.

51. Reyniers, J. P., Woodard, J. C. and Alvarez, M. R. (1974). Nuclear cytochemical alterations in α-protein induced nephrocytomegalia. Lab. Invest. 30, 582-588.

52. Woodard, J. C. (1969). On the pathogenesis of alpha protein-induced nephrocytomegalia. Lab. Invest. 20, 9-16.

53. Newberne, P. M. and Young, V. R. (1966). Effects of diets marginal in methionine and choline with and without vitamin B_{12} on rat liver and kidney. J. Nutr. 89, 69-79.

54. Struthers, B. J., Dahlgren, R. R. and Hopkins, D. T. 1976. Biological effects of feeding graded levels of alkali hydrolyzed soy protein containing lysinoalanine in Sprague-Dawley and Wistar rats. J. Nutr., 107, 1190-1199.

55. Karayiannis, N. (1976). Lysinoalanine Formation in Alkali Treated Proteins and their Biological Effects, Ph.D. Thesis. University of California, Berkeley.

56. O'Donovan, C. J. (1976). Recent studies of lysinoalanine in alkali-treated proteins. Fd. Cosmet. Toxicol. 14, 483-489.

57. Feron, V. J., van Beek, L., Slump, P. and Beems, R. B. (1977). Toxicological aspects of alkali treatment of food proteins. In: "Biochemical Aspects of New Protein Food", J. Adler Nissen, ed.), Pergamon Press, Oxford, England, pp. 139-147.

58. Karayiannis, N. I., MacGregor, J. R., and Bjeldanes, L. F. (1979). Biological effects of alkali-treated soy protein and lactalbumin in the rat and mouse. Food and Cosm. Toxicol., in press.

59. Friedman, M. (1977). Crosslinking amino acids—stereochemistry and nomenclature. In: "Protein Crosslinking: Nutritional and Medical Consequences", (M. Friedman, ed.), Plenum Press, New York, pp. 1-27.

60. Evans, G. W. and Hahn, C. J. (1974). Copper- and zinc-binding components in rat intestine. In: "Protein-Metal Interactions," (M. Friedman, ed.), Plenum Press, New York, pp. 285-297.

61. MacGregor, J. T. and Clarkson, T. W. (1974). Distribution, tissue binding and toxicity of mercurials. In: "Protein-Metal Interactions", (M. Friedman, ed.), Plenum Press, New York, pp. 463-503.

62. Finot, P.-A., Bujard, E., and Arnaud, M. (1977). Metabolic transit of lysinoalanine (LAL) bound to protein and of free radioactive [^{14}C]-lysinoalanine. In: "Protein Crosslinking — Nutritional and Medical Consequences," M. Friedman, Ed., Plenum Press, N.Y., pp. 51-71.

63. Sternberg, M., Kim, C. Y., and Schwende, F. J. (1975). Lysinoalanine: presence in foods and food ingredients. Science 190, 992-994.

64. Sternberg, M. and Kim, C. Y. (1977). Lysinoalanine formation in protein food ingredients. In: "Protein Crosslinking: Nutritional and Medical Consequences", M. Friedman, Ed., Plenum Press, New York, pp. 73-84.

65. Erbersdobler, H. F., von Wangenheim, B., Hänichen, T. Adverse effects of Maillard products — especially of fructoselysine in the organism. Paper presented at XI Internat. Congress of Nutrition, Rio de Janeiro, Brazil, Aug. 1978.

66. Zumberge, J. E. The effect of D-glucose on aspartic acid racemization. Paper presented at the Carnegie Institution of Washington Conference: Advances in the Biogeochemistry of Amino Acids, Warrenton, Virginia, Oct. 29–Nov. 1, 1978.

67. Berg, C. P. (1959). Utilization of D-amino acids. In: "Protein and Amino Acid Nutrition," A. A. Albanese, Ed., Academic Press, N. Y., PP. 57–96.

68. Gibson, Q. H. and Wiseman, G. (1951). Selective absorption of stereoisomers of amino acids from loops of the small intestine of the rat. Biochem. J. 48: 426–429.

69. Finch, L. R. and Hird, F.J.R. (1960). The uptake of amino acids by isolated segments of rat intestine. II. A survey of affinity for uptake from rates of uptake and competition of uptake. Biochim. Biophys. Acta 43, 278–287.

70. Rosenhagen, M. and Segal, S. (1974). Stereospecificity of amino acid uptake by rat and human kidney cortex slices. Am. J. Physiol. 227, 843–847.

71. Meister, A. (1965). "Biochemistry of the Amino Acids," Vol. I, Academic Press, N.Y., pp. 338–369.

72. Rose, W. C. 1949. Amino acid requirements of man. Fed. Proc. 8, 546–552.

73. Kies, C., Fox, H. and Aprahamian, S. (1975). Comparative value of L-, DL-, and D-methionine supplementation of an oat-based diet for humans. J. Nutr. 105, 809–814.

74. Zezulka, A. Y. and Calloway, D. H. (1976). Nitrogen retention in men fed isolated soybean protein supplemented with L-methionine, D-methionine, N-acetyl-L-methionine, or inorganic sulfate. J. Nutr. 106, 1286–1291.

75. Wretlind, J.A.J. (1952). The effect of D-amino acids on the stereonaturalization of D-methionine. Acta Physiol. Scand. 25, 267–275.

76. Van Beek, L., Feron, V. J. and de Groot, A. P. (1974). Nutritional effects of alkali-treated soy protein in rats. J. Nutr. 104, 1630–1636.

77. Efron, M. L., McPherson, T. C., Shih, V. E., Welsh, C. F. and MacCready, R. A. (1969). D-Methioninuria due to DL-methionine ingestion. Am. J. Dis. Child. 117, 104–107.

78. Chem. Week, (Oct. 4, 1978). They're moving in for the kill against pain, p. 47-48.

79. Munro, H. N. (1978). Nutritional consequences of excess amino acid intake. In: "Nutritional Improvement of Food and Feed Proteins", M. Friedman, Ed., Plenum Press, New York, pp. 119-129.

80. Johnson, R. C. and Shah, S. N. (1979). Effects of phenylalanine and cortisol treatments on brain development and myelination in suckling rats. Fed. Proc., 38(3), 716.

81. Stegink, L. D., Filer, L. J. R., Baker, G. L. and McDonnel, J. E. (1979). Effect of Aspartame loading upon plasma and erythrocyte amino acid levels in phenylketonuric heterozygotes and normal adult subjects. J. Nutr. 109, 708-717.

82. Friedman, M. and Krull, L. H. (1970). N- and C-Alkylation of proteins in dimethyl sulfoxide. Biochim. Biophys. Acta, 207, 361-363.

83. Cram, D. J. (1965). "Fundamentals of Carbanion Chemistry", Academic Press, New York and London. Chapter 4.

RECEIVED October 18, 1979.

Deterioration of Food Proteins by Binding Unwanted Compounds Such as Flavors, Lipids and Pigments

SOICHI ARAI

Department of Agricultural Chemistry, University of Tokyo,
Bunkyo-ku, Tokyo 113, Japan

Proteins often interact with other compounds found in biological materials. The retention and stability of a flavor
generally is markedly improved in the presence of protein. This
is the result of some type of protein-flavor interaction which
decreases the volatility of the flavor. When a flavor is an undesirable one, such interaction may cause the protein to be
organoleptically unacceptable as human food. It is often a very
formidable task to remove a protein-bound flavor completely in
order to prepare a bland protein. Many lipids and natural pigments, physiologically significant to living systems, are
important in foods. While they often provide desirable color and
texture to food, both lipids and pigments may cause undesirable
color changes and deteriorative changes in proteins. In a few
cases toxic constituents may be formed. Therefore, interaction
of constituents in foods may be both boon and bane (1).

The present paper reviews some of the undesirable effects
resulting from the interaction of flavor constituents, lipids and
pigments with proteins. Our laboratory as well as others have
contributed to this knowledge.

Interaction of Proteins with Flavors

Examples of interaction of protein with both volatile and
non-volatile flavor constituents are available. One example is
the interaction between gelatin and several non-volatile flavor
nucleotides: 5'-GMP, 5'-IMP, 5'-AMP and 5'-CMP. Saint-Hilaire
and Solms (2) equilibrated solutions of 5 - 90 mM nucleotide in
0.004 % gelatin at pH 6.5 and determined bound nucleotide by
ultraviolet spectroscopy. They analyzed the results by use of
the Scatchard equation:

$$\bar{r} \, / \, (n - \bar{r}) = KC \tag{1}$$

where $\bar{r}$ is the average number of moles of bound ligand per mole of
protein, n is the maximum number of moles of ligand bound per mole

0-8412-0543-4/80/47-123-195$05.00/0
© 1980 American Chemical Society

of protein, K is a binding constant and C is the molar concentration of non-bound ligand at equilibrium (3). The potent flavor
potentiators, 5'-GMP and 5'-IMP, gave K values of 235 and 112
mM^{-1}, respectively, while 5'-AMP and 5'-CMP, known to have poor
flavor-potentiating activities, had much smaller K values. No
other reports of protein interaction with such non-volatile
flavors are available.

More attention has been given to interaction of proteins with
volatile flavors, especially with volatile carbonyls. Nawar (4)
found that addition of gelatin to solutions of a homologous series
of 2-alkanones caused decreases in their volatilities. Hawrysh
and Stine (5) reported on the retention of 2-alkanones in a model
system that simulated blue-vein cheese. A similar but more systematic experiment was carried out by Franzen and Kinsella (6).
By headspace analysis via gas chromatography, they quantified the
retention of a variety of volatile aldehydes and ketones by food
proteins such as α-lactalbumin, bovine serum albumin, leaf protein
concentrate, single-cell protein and soy protein preparations.
Although the quantity of retained flavor depended on the type,
amount and composition of protein as well as on the presence of
lipids, it was clear that protein decreased to some extent the
volatilities of flavors by adsorbing or occluding them. For example, the addition of protein to an aqueous system containing 1-
hexanal caused a 9 - 23 % decrease in its volatility. Gremli (7)
also determined the headspace composition of a model system of a
10:1 mixture of soy protein and aldehyde in water. The percent
retention of aldehydes were as follows: 1-hexanal, 37 - 44 %; 1-
heptanal, 62 - 72 %; 1-octanal, 83 - 85 %; 1-nonanal, 90 - 93 %; 1-
decanal, 94 - 97 %; 1-undecanal, 96 - 100 %; 1-dodecanal, 94 - 100 %;
2-hexenal, 68 - 75 %; 2-heptenal, 82 - 88 %; 2,6-nonadienal, 90 - 98
%; 2,4-nonadienal, 92 - 97 %; 2-decenal, 100 %; and 2-dodecenal,
100 %. In these experiments, the last two volatile compounds
behaved as if they were completely non-volatile. Beyeler and
Solms (8) calculated the binding constant (K) for several volatile
compounds in the presence of soy protein and bovine serum albumin
by the following equation:

$$\bar{r} = KC \qquad\qquad\qquad\qquad (2)$$

where C is the molar concentration of free ligand at equilibrium
and $\bar{r}$ is the average number of moles bound ligand per mole of
protein. The data showed that the K values for aldehydes are
generally larger than those for other classes of compounds.

Strong interaction of volatile aldehydes occur naturally in
soy protein products (9). Arai et al. (10) found that 1-hexanal
is one of the major odorants of soybean and that this aldehyde
interacts readily with soy protein. In order to determine
whether the interaction is enhanced by denaturation of protein,
Arai et al. (9) did three experiments under different conditions.
In the first experiment (I), an acid-precipitated fraction of

native soy protein (10 g) was dissolved in water (100 ml) and
various amounts of 1-hexanal were added. The resulting mixtures
were vigorously agitated at 20°C for 5 h in a sealed flask filled
with nitrogen. Each sample was then freeze-dried. In the second
experiment (II), various amounts of 1-hexanal were added to the
acid-precipitated protein (10 g) dissolved in water (100 ml). The
mixtures, in flasks equipped with condensers, were heated at 90°C
for 1 h with vigorous stirring and then freeze-dried. In the
third experiment (III), identical hexanal/protein mixtures to
those above were heated at 90°C for 24 h under reflux conditions
and then lyophilized. Aliquots of the freeze-dried samples were
dissolved in a NaOH solution (final pH ca. 13) to liberate the 1-
hexanal bound by protein. Gas chromatography was used to determine the liberated 1-hexanal. Figure 1 shows the quantity of 1-
hexanal bound depended upon the heat treatment of protein as well
as upon the amount of aldehyde initially added.

Arai et al. (9) obtained a binding constant (K) for the
interaction of 1-hexanal with the partially denatured soy protein
(Experiment II above) from gel filtration data analyzed by using
Beidler's equation:

$$n \; / \; (S - n) = KC \tag{3}$$

where C is the concentration of total ligand, n is the amount of
bound ligand when the initial total ligand concentration equals C,
and S is the amount when the ligand concentration has reached a
maximum (11). The analysis gave $K = 173 \; M^{-1}$ and $S = 0.847$ mg/g protein. The S value indicates that the partially denatured soy protein bound 1-hexanal to almost 0.1 % of its weight. An additional
study (9) demonstrated that hydrolysis of the protein decreased
the 1-hexanal binding ability (Table I). The amount of 1-hexanal
liberated by the enzyme treatment correlated well with the degree
of hydrolysis of protein (Table I). Arai et al. (12) have also
reported that free tryptophan in an enzymatic hydrolysate of soy
protein reacts with 1-hexanal to form the condensation products,
1-pentyl-2,3,4,9-tetrahydro-1H-pyrido[3,4-b]indole-3-carboxylic
acid (I), 1-pentyl-4,9-dihydro-3H-pyrido[3,4-b]indole-3-carboxylic
acid (II) and 1-pentyl-9H-pyrido[3,4-b]indole (III). Related compounds are discussed later with respect to potent mutagenicity.

Volatile aldehydes, including 1-hexanal, may be primarily
responsible for the beany odor of soybean (10, 13, 14). They are
present even in defatted soybean flour. Recently, Chiba et al.
(15) have deodorized soybean flour by treatment with aldehyde
dehydrogenase from bovine liver. Deodorization was a result of
converting aldehydes to acids, e.g., 1-hexanal to caproic acid.
They postulated that both free and bound aldehydes can act as substrates for this dehydrogenase. Consequently, enzymatic treatment
resulted in a product without beany odor.

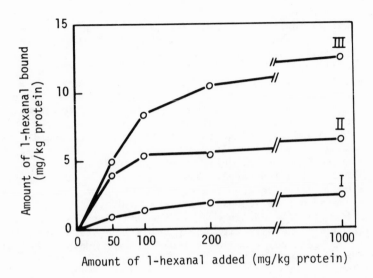

Agricultural and Biological Chemistry

Figure 1. Binding of 1-hexanal by soy protein. I, II and III refer to the first, second and third experiments, respectively. For details see text (9).

Table I. Relationship between the degree of hydrolysis
of soy protein* and the amount of 1-hexanal retained

Degree of hydrolysis** (%)	Amount of 1-hexanal retained (mg / kg hydrolysate)
0	6.67
22.1	5.20
41.6	2.41
72.7	1.03
99.8	0.05

* Partially denatured soy protein (see text).
** Hydrolysis was performed at pH 2.8 with a microbial
acid protease (Molsin). Each degree of hydrolysis was
measured with 10 % trichloroacetic acid (TCA) and represented as (TCA-soluble N / total N) × 100.
    From Arai et al. (9)

Agricultural and Biological Chemistry

(4)

Interaction of Protein with Lipids

Lipids are often a nuisance in extraction of proteins. For example, in preparing leaf protein concentrate a protein-lipid complex is formed frequently affecting the protein extraction efficiency (16). Nutritionally, the complex is disadvantageous because it resists digestion by proteases (17). Shenouda and Pigott (18) found that the formation of protein-bound lipids can cause a low efficiency of extraction of protein from fish. Hydrophobic bonding probably plays an important role in protein-lipid interactions. Mohammadzadeh-k et al. (19) reported protein interaction with completely apolar compounds such as aliphatic hydrocarbons.

Shenouda and Pigott (20) have studied the interaction of polar lipids with protein. Using a spin-label technique, they demonstrated that heat-denatured myosin from fish muscle bound polar lipids more tightly than neutral triglycerides. They showed that a solvent with higher polarity permits better separation of lipids from protein (21). Similar examples of the interaction of polar lipids with proteins have been found in peas (22) and in soy protein after heat-denaturation (23). According to Noguchi et al. (24), soy protein curd, prepared by heat-denaturation of protein followed by salting-out, contains a variety of polar lipids including phosphatidylcholine, phosphatidylethanolamine and phosphatidylinositol. They found that a significant amount of these lipids is probably in a protein-bound state resistant to extraction with chloroform/methanol. When the curd was incubated with an acid protease (Molsin), most of the bound phospholipids were liberated to an extent dependent on the incubation time.

Ohtsuru et al. (25) have recently investigated the behavior of phosphatidylcholine in a model system that simulated soy milk. They used spin-labelled phosphatidylcholine (PC*) synthesized from egg lysolecithin and 12-nitroxide stearic acid anhydride. The ESR spectrum of a mixture of PC* (250 µg) and native soy protein (20 mg) homogenized in water by sonication resembled that observed for PC* alone before sonication. However, when PC* (250 µg) was sonicated in the presence of heat-denatured soy protein (20 mg), splitting of the ESR signal occurred. On this basis, they postulated the existence of two phases: PC* making up a fluid lamella phase and PC* immobilized probably due to the hydrophobic interaction with the denatured protein. In a study of a soy-milk model, Ohtsuru et al. (25) reported that a ternary protein-oil-PC* complex occurred when the three materials were subjected to sonication under the proper condition. Based on data from the ESR study, a schematic model has been proposed for the reversible formation-deformation of the ternary complex in soy milk (Figure 2).

Serious problems arise when protein-bound lipids undergo autoxidation followed by decomposition. Castell (26) showed that formaldehyde formed by the decomposition of autoxidized fish oil can cause toughened texture of fish protein. St. Angelo and Ory

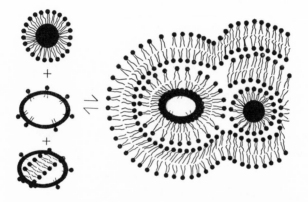

Figure 2.   A model for the formation-deformation of a ternary protein-oil-phos-phatidylcholine complex.   The protein molecule is represented as a large open circle and the phosphatidylcholine molecule as a small filled circle having two tails.   The large filled circle represents an oil particle (25).

(27) reported on the undesirable effects of unsaturated lipids on protein; their autoxidation produced aldehydes and ketones which reacted with some amino acid residues. Roubal and Tappel (28) and Roubal (29) showed that protein deterioration could be initiated by free radicals originating from lipid peroxidation. Peroxidation resulted in oxidation of some amino acid residues as well as cross-linking of proteins via malonaldehyde formed.

Karel's group at M.I.T. has also studied the effect of peroxidizing methyl linoleate in the presence of lysozyme and other selected food proteins in model systems (30 - 33). ESR studies indicated that protein free radicals were formed as a result of the interaction, especially when the reacting system was maintained at a water activity of 0.75. The free radicals primarily showed central singlet lines, attributable to carbon-centered radicals with $g = 2.004$ and guanidyl nitrogen-centered radicals with $g = 2.0027$. Substantial evidence indicated formation of free radicals with the side-chains of tryptophan, histidine, lysine and arginine residues of protein. Proteins containing cysteine also exhibited downfield shoulders at $g = 2.015$ and 2.023 that were essentially identical to peaks obtained with free cysteine. Based on these data, Karel and coworkers postulated that free radical transfer to protein occurs via complex formation of the following type:

$$PH + LOOH \rightarrow [PH \cdots LOOH] \rightarrow P \cdot + LOO \cdot + H_2O \qquad (5)$$

where PH and LOOH refer to protein and lipid hydroperoxide, respectively. An intermolecular reaction of $P \cdot$ may then take place leading to protein polymerization of the type $P-(P)_n-P \cdot$. It was shown that lysozyme polymerized (covalently) during incubation with peroxidizing methyl linoleate, with a decrease in solubility as well as in enzyme activity. Such a peroxide-initiated radical reaction can cause chemical deterioration of wheat gliadin, bovine serum albumin and ovalbumin as well. Free radicals trapped in matrices of proteins, particularly of denatured proteins, can be kept stabilized over a long period of time (29). Undoubtedly, food proteins undergo radical-induced damage during their prolonged storage in the presence of unsaturated lipids.

Interaction of Proteins with Pigments and Related Compounds

Synthetic dyes can interact with proteins. Experiments have been carried out to assess binding capacities of proteins toward a variety of food dyes (34 - 36). These include azo dyes (new coccine, amaranth, orange G, etc.), triphenylmethane dyes (guinea green, brilliant blue FCF, acid violet 6B, etc.) and isoxanthene dyes (rose bengal, erythrosine, eosine, etc.). An example of physiological importance is the work of Tokuma and Terayama (37). They identified alcohol dehydrogenase as a target protein for binding of a carcinogenic aminoazo dye. It is, however, beyond

the scope of this article to deal with these topics; therefore the
discussion will concentrate on the interaction of proteins with
naturally occurring pigments and related compounds.

Special problems arise during extraction of proteins from
green leaves and algae. Much effort has been devoted to separation of chlorophylls, carotenoids and other photosynthetic pigments from leaf proteins (38 - 44). A tight chlorophyll-protein
complex formed, for example, during greening of etiolated bean
leaves is a nuisance (45). A similar example is chlorophyll-
lipoprotein complex formation in chloroplasts of Chlorella (46).
Feeding tests using rats showed that the complex was primarily
responsible for the low nutritive value of the Chlorella protein.
Arai et al. (47) attempted to extract an alkali-soluble protein
from the blue-green alga, Spirulina maxima, and another one from
the non-sulfur purple bacterium, Rhodopseudomonas capsulatus. In
both cases, pretreatment of dry cells with ethanol removed most of
the photosynthetic pigments. However, the extracted proteins were
still brown. Gel chromatography on Sephadex G-15 showed that the
remaining pigment was tightly bound to protein. The pigment could
be separated from pepsin-treated protein on a Sephadex column.
Algal pigments include the so-called biliproteins such as phyco-
erythrins and phycocyanins which are covalently bound to protein
(48).

Higher plants contain a chromoprotein called phytochrome,
which occurs in two interconvertible forms, P_R and P_{FR}, with absorption maxima at 665 nm and 725 nm, respectively. Walker and
Bailey (49) believe that the interconvertible photoreaction of
phytochrome may be involved in the photoregulation of growth,
development, adaptation and other functions of higher plants.
Rüdiger (50) proposed a model of the two forms of phytochrome
involving covalently bound linear tetrapyrrole. The chromophore
is a bile pigment. Fry and Mumford (51) have isolated a chromo-
phore-containing peptide by digesting phytochrome with protease;
therefore it may be possible to use proteolysis to depigment the
chromoprotein.

Besides the above-mentioned pigments, a variety of closely
related colorless compounds occur widely in plants. Among these
are polyphenols, some of which have a potent tanning activity.
Because of their specific affinity for proteins, tannins have been
used in protein analysis. The method can be improved further by
characterizing the mode of protein-tannin interaction (52) and by
optimizing the conditions for quantitative analysis (53, 54).
Tannins often cause nutritional deficiencies and toxicities as
shown for the protein extracted from high-tannin species of
sorghum (55). Another example is the inhibitory effect of oak
leaf tannin on the tryptic digestion of proteins (56). Chlorogenic
acid, a ubiquitous depside-type tannin, also can associate tightly
with leaf proteins, affecting their tryptic digestibility (57,58).
Dryden and Satterlee (59) showed that chlorogenic acid added to
casein reacts covalently to prevent the protein from undergoing

in vivo digestion. The growth of Tetrahymena on the chlorogenic
acid-treated casein was decreased. Other polyphenols also react
with proteins, for example, with arachin (60).

Phenolic compounds including flavonols, although colorless,
can be enzymatically and/or non-enzymatically oxidized to pigments. When such oxidation takes place in the presence of protein,
the protein may become pigmented. Igarashi et al. (61) measured
the effect of a number of phenolic compounds on browning of casein
solution at pH 6.8 (Table II). The results indicate that the
position of the OH groups of the phenolic compounds determines the
color intensity. In particular, 3- and 7-OH groups on the flavone
ring bearing 3'- and 4'-OH groups appear to be required for
maximum browning of the casein solution. From a nutritional point
of view it should be noted that available lysine of casein
decreased in accordance with the brown color intensity. Igarashi
et al. (62) also showed that in vitro digestibility of casein
decreased when the protein was incubated with quercetin. In this
case a significant degree of damage occurred to methionine as well
as to lysine.

Horigome and Kandatsu (63) prepared a Fuki (Petasites japon-
icus miq.; a traditional food plant in Japan) acetone-powder with
a high o-diphenol oxidase activity. When the powder was added to
a mixture of caffeic acid and casein, casein was gradually pigmented. A feeding test with rats demonstrated that there was a
close relationship between the decrease in biological value of the
pigmented casein and its color intensity.

Finally, examples of the effect of protein interaction with
fluorescent compounds are available. Lohrey et al. (64) demonstrated that a photosensitizing effect (photodynamic effect)
appeared when rats were fed on a diet containing leaf protein concentrate prepared from lucerne (Medicago sativa). Skin lesions of
varying severity up to the sloughing of ears and tails occurred
when the fed rats were illuminated with natural daylight through
window glass. Extracts from blood plasma and livers of rats given
the leaf protein concentrate contained phaeophorbide-a which is a
fluorescent compound derived from chlorophyll-a by removal of
magnesium and phytol. This finding is supported by the work of
Isobe et al. (65) who have clearly shown that phaeophorbide-a does
cause hypersensitization in rats.

Much more attention is currently being paid to carcinogenicity of fluorescent compounds. A typical example may be aflatoxin (66). Beckwith et al. (67) have studied the interaction
between corn protein and aflatoxin B_1. Tryptophan derivatives
also are of interest because of their possible carcinogenicity.
Sugimura et al. (68) identified the following two mutagenic principles: 3-amino-1,4-dimethyl-5H-pyrido[4,3-b]indole (I) and
3-amino-1-methyl-5H-pyrido[4,3-b]indole (II) in a pyrolyzate of
tryptophan. Subsequently, Yoshida et al. (69), in investigations
on factors inducing mutagenicity in Salmonella typhimurium TA 98,
identified two related compounds: 2-amino-9H-pyrido[2,3-b]indole

Table II. Effect of phenolic compounds added to a
casein solution on its browning during incubation*

Phenolic compounds	Optical density	
	at 470 nm	at 520 nm
Chlorogenic acid	0.521	0.396
Caffeic acid	0.751	0.413
Catechol	0.850	0.510
Quercetin	0.941	0.724
Kaempherol	0.206	0.174
Myricetin	1.505	1.007
Dihydroquercetin	0.590	—
Protocatechuic acid	0.314	0.260
Phloroglucinol	0.780	0.331
Azaleatin	0.778	0.510
Rhamnetin	0.206	0.174
Quercitrin	0.078	0.055
Rutin	0.047	0.033
Luteolin	0.196	0.174

* Casein (1 g) was dissolved in 100 ml of 0.1 M
phosphate (pH 6.8) containing 0.1 mM phenolic compound. The solution was refluxed at 80°C for 10 h
under aeration prior to measurement of optical
densities.
    From Igarashi et al. (61)

Journal of the Agricultural Chemical Society of Japan

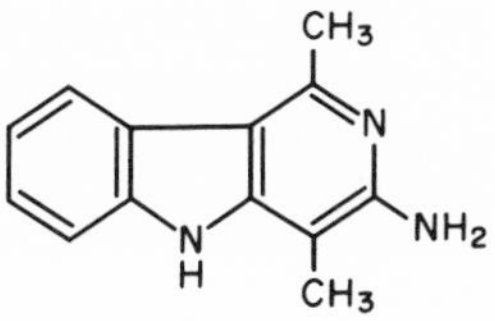

I

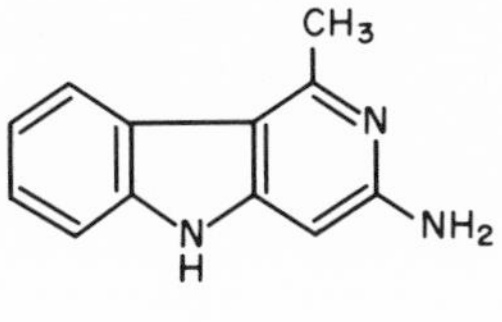

II

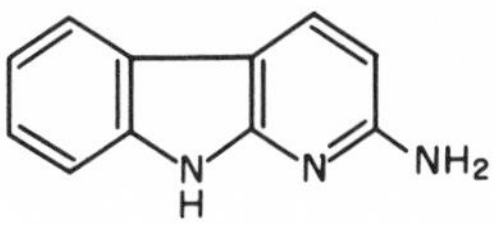

III

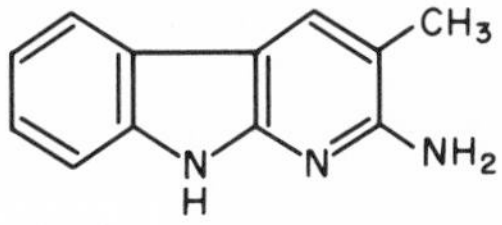

IV

(III) and 2-amino-3-methyl-9H-pyrido[2,3-b]indole (IV) in a pyrolyzate of soy protein. These fluorescent compounds could be
protein-bound in roasted foods. Detailed experiments on the
interaction between food proteins and such compounds of toxicological importance are being carried out in Japan.

Discussion

Proteins can bind flavor constituents, especially volatile
carbonyls. When bound to protein, some volatile aldehydes behave
as if they were non-volatile compounds and are retained over a
long period of time during storage. Such interaction may cause
chemical as well as organoleptical deterioration of food proteins.
As exemplified by soy protein, interaction of protein with
1-hexanal is promoted by heat denaturation. The interaction is
decreased by treating a heat-denatured soy protein with protease;
the aldehyde is liberated to an appreciable extent.
Lipids can act as precursors of a variety of aliphatic carbonyls with objectionable flavors. Proteins interact with lipids
as well. The interaction is primarily through hydrophobic bonding.
Additionally, a type of polar interaction may be involved,
particularly when phospholipids take part. In the case of soy
milk, a ternary protein-oil-phosphatidylcholine complex probably
occurs.
A serious problem arises when lipids undergo peroxidation in
the presence of protein. Free radicals originating from hydroperoxides transfer to several specific amino acid residues of protein.
As a result, the protein undergoes radical induced polymerization,
with loss of solubility and other original properties including
nutritive value.
Naturally occurring pigments also are bound by proteins to a
greater or lesser extent depending on chemical structure. Photosynthetic pigments often obstruct the process for preparing pure
leaf protein concentrate. A chlorophyll-lipoprotein complex
occurring in the algal chloroplast fraction is a nuisance in that
it resists in vivo as well as in vitro digestion. Of physiological
and toxicological importance is the possibility that some fluorescent compounds with photosensitizing and mutagenic activities
could remain in foods in a protein-bound state.
Besides pigments, closely related compounds occur naturally.
Among these are phenolic compounds. Chlorogenic acid, a ubiquitous compound, interacts readily with proteins, affecting their
digestibility. Many other common phenolic compounds, although
colorless, undergo enzymatic and/or non-enzymatic oxidation to
pigments. When such pigmentation takes place with proteins,
deteriorative changes result.
Examples given in this chapter suggest caution in the use of
some proteins for food. A great deal of time and effort has been
spent in attempting to remove flavors, lipids, pigments, etc. from
proteins. Their treatment with proteases may be generally useful

tool for this purpose. Use of non-protease enzymes also seems
promising, although further studies are needed to provide
information applicable in the industrial production of wholesome
proteins for human consumption.

Literature Cited

1. Feeney, R. E., in "Evaluation of Proteins for Humans",
    Bodwell, C. E., Ed.; Avi Publ. Co., 1977, p. 233.
2. Saint-Hilaire, P.; Solms, J., J. Agric. Food Chem., 1973, 21,
    1126.
3. Scatchard, G., Ann. N.Y. Acad. Sci., 1949, 51, 660.
4. Nawar, W. W., J. Agric. Food Chem., 1971, 19, 1057.
5. Hawrysh, Z. J.; Stine, C. M., J. Food Sci., 1973, 38, 7.
6. Franzen, K. L.; Kinsella, J. E., J. Agric. Food Chem., 1974,
    22, 675.
7. Gremli, H. A., J. Amer. Oil Chem. Soc., 1974, 51, 95A.
8. Beyeler, M.; Solms, J., Lebensm.-Wiss. u. Technol., 1974, 7,
    217.
9. Arai, S.; Noguchi, M.; Yamashita, M.; Kato, H.; Fujimaki, M.,
    Agric. Biol. Chem. Japan, 1970, 34, 1569.
10. Arai, S.; Noguchi, M.; Kaji, M.; Kato, H.; Fujimaki, M.,
    Agric. Biol. Chem. Japan, 1970, 34, 1420.
11. Beidler, L. M., J. Gen. Physiol., 1954, 38, 133.
12. Arai, S.; Abe, M.; Yamashita, M.; Kato, H.; Fujimaki, M.,
    Agric. Biol. Chem. Japan, 1971, 35, 552.
13. Qvist, I. H.; Von Sydow, E. C. F., J. Agric. Food Chem.,
    1974, 22, 1077.
14. Wolf, W. J., J. Agric. Food Chem., 1975, 23, 136.
15. Chiba, H.; Takahashi, N.; Yoshikawa, M.; Sasaki, R.,
    Abstracts of Papers, p. 509, Annual Meeting of the
    Agricultural Chemical Society of Japan, Nagoya, 1978.
16. Buchanan, R. A., J. Sci. Food Agr., 1969, 20, 359.
17. Buchanan, R. A., Br. J. Nutr., 1969, 23, 533.
18. Shenouda, S. Y. K.; Pigott, G. M., J. Food Sci., 1975, 40,
    523.
19. Mohammadzadeh-k, A.; Feeney, R. E.; Samuels, R. B.; Smith,
    L. M., Biochim. Biophys. Acta, 1967, 147, 583.
20. Shenouda, S. Y. K.; Pigott, G. M., J. Food Sci., 1974, 39,
    726.
21. Shenouda, S. Y. K.; Pigott, G. M., J. Food Sci., 1975, 40,
    520.
22. Hayder, M.; Hadziyev, D., J. Food Sci., 1973, 38, 772.
23. Kamat, V. B.; Graham, G. E.; Davis, M. A. F., Cereal Chem.,
    1978, 55, 295.
24. Noguchi, M.; Arai, S.; Kato, H.; Fujimaki, M., J. Food Sci.,
    1970, 35, 211.
25. Ohtsuru, M.; Kito, M.; Takeuchi, Y., Agric. Biol. Chem.
    Japan, 1976, 40, 2261.
26. Castell, C. H., J. Amer. Oil Chem. Soc., 1971, 48, 645.

27. St. Angelo, A. J.; Ory, R. L., J. Agric. Food Chem., 1975, 23, 141.
28. Roubal, W. T.; Tappel, A. L., Arch. Biochem. Biophys., 1966, 113, 150.
29. Roubal, W. T., J. Amer. Oil Chem. Soc., 1970, 47, 141.
30. Karel, M.; Schaich, K.; Roy, R. B., J. Agric. Food Chem., 1975, 23, 159.
31. Schaich, K. M.; Karel, M., J. Food Sci., 1975, 40, 456.
32. Kanner, J.; Karel, M., J. Agric. Food Chem., 1976, 24, 468.
33. Schaich, K. M.; Karel, M., Lipids, 1976, 11, 392.
34. Aizawa, H., J. Japanese Soc. Food Nutr., 1969, 22, 231.
35. Aizawa, H., J. Japanese Soc. Food Nutr., 1969, 22, 240.
36. Aizawa, H.; Takeyama, I., J. Japanese Soc. Food Nutr., 1969, 22, 235.
37. Tokuma, Y.; Terayama, H., Biochem. Biophys. Res. Commun., 1973, 54, 341.
38. Arkcoll, D. B., J. Sci. Food Agr., 1969, 20, 600.
39. Arkcoll, D. B., J. Sci. Food Agr., 1973, 24, 437.
40. Arkcoll, D. B.; Holden, M., J. Sci. Food Agr., 1973, 24, 1217.
41. Knuckles, B. E.; de Fremery, D.; Bickoff, E. M.; Kohler, G. O., J. Agric. Food Chem., 1975, 23, 209.
42. Igarashi, K.; Sakamoto, Y.; Yasui, T., J. Agric. Chem. Soc. Japan, 1976, 50, 67.
43. Little, A. C., J. Food Sci., 1977, 42, 1570.
44. Bray, W. J.; Humphries, C.; Ineritei, M. S., J. Sci. Food Agr., 1978, 29, 165.
45. Argyroudi-Akoyunoglou, J. H.; Feleki, Z.; Akoyunoglou, G., Biochem. Biophys. Res. Commun., 1971, 45, 606.
46. Ishii, T.; Kandatsu, M.; Kametaka, M., J. Japanese Soc. Food Nutr., 1974, 27, 103.
47. Arai, S.; Yamashita, M.; Fujimaki, M., J. Nutr. Sci. Vitaminol., 1976, 22, 447.
48. Ó hEocha, C., in "Biochemistry of Chloroplasts", Vol. I, Goodwin, T. W., Ed.; Academic Press, 1966, p.411.
49. Walker, T. S.; Bailey, J. L., Biochem. J., 1970, 120, 613.
50. Rüdiger, W., Liebig's Ann. Chem., 1969, 723, 208.
51. Fry, K. T.; Mumford, F. E., Biochem. Biophys. Res. Commun., 1971, 45, 1466.
52. Van Buren, J. P.; Robinson, W. B., J. Agric. Food Chem., 1969, 17, 772.
53. Hoff, J. E.; Singleton, K. I., J. Food Sci., 1977, 42, 1566.
54. Hagerman, A. E.; Butler, L. G., J. Agric. Food Chem., 1978, 26, 809.
55. Guiragossian, V.; Chibber, B. A. K.; Van Scoyoc, S.; Jambunathan, R.; Mertz, E. T.; Axtell, J. D., J. Agric. Food Chem., 1978, 26, 219.
56. Feeney, P. P., Phytochem., 1969, 8, 2119.
57. Lahiry, N. L.; Satterlee, L. D., J. Food Sci., 1975, 40, 1326.
58. Lahiry, N. L.; Satterlee, L. D.; Hsu, H. W.; Wallace, G. W., J. Food Sci., 1977, 42, 83.

59. Dryden, M. J.; Satterlee, L. D., J. Food Sci., 1978, 43, 650.
60. Neucere, N. J.; Jacks, T. J.; Sumrell, G., J. Agric. Food
    Chem., 1978, 26, 214.
61. Igarashi, K.; Shishido, N.; Yasui, T., J. Agric. Chem. Soc.
    Japan, 1978, 52, 499.
62. Igarashi, K.; Suzuki, M.; Majima, T.; Yasui, T., J. Agric.
    Chem. Soc. Japan, 1978, 52, 219.
63. Horigome, T.; Kandatsu, M., J. Japanese Soc. Food Nutr.,
    1971, 24, 253.
64. Lohrey, E.; Tapper, B.; Hove, E. L., Br. J. Nutr., 1974, 31,
    159.
65. Isobe, A.; Sasaki, R.; Kimura, S., J. Japanese Soc. Food
    Nutr., 1977, 30, 99.
66. Butler, W. H., in "Aflatoxin", Goldblatt, L. A., Ed.;
    Academic Press, 1969, p. 223.
67. Beckwith, A. C.; Vesonder, R. F.; Ciegler, A., J. Agric. Food
    Chem., 1975, 23, 582.
68. Sugimura, T.; Kawachi, T.; Nagao, M.; Yahagi, T.; Seino, Y.;
    Okamoto, T.; Shudo, K.; Kosuge, T.; Tsuji, K.;
    Wakabayashi, K.; Iitaka, Y.; Itai, A., Proc. Japan
    Academy, 1977, 53, 58.
69. Yoshida, D.; Matsumoto, T.; Yoshimura, R.; Matsuzaki, T.,
    Biochem. Biophys. Res. Commun., 1978, 83, 915.

RECEIVED October 18, 1979.

Deteriorative Changes of Proteins During Soybean Food Processing and Their Use in Foods

DANJI FUKUSHIMA

Kikkoman Foods, Inc., Walworth, WI 53184

Various deteriorative changes occur in proteins during food processing and food storage, even under mild conditions. However, a deteriorative change for one purpose can be a favorable one for another purpose. For instance, some meat proteins change physical properties during frozen storage, resulting in loss of chewing qualities and/or functional properties such as binding or emulsifying properties (1). Therefore, this change during frozen storage is a deteriorative one for meats. However, this change is an advantage for the manufacture of a soybean protein product known as "kori-tofu" described later. Another example where a deteriorative change from one aspect can be a favorable one from another aspect is the insolubilization of soybean protein during evaporation (2,3). This insolubilization of proteins is a deteriorative one for the manufacture of "yuba", another soybean protein product described later.

The deterioration of physical properties of proteins during food processing and/or storage described above are due to irreversible insolubilization of the proteins. Irreversible insolubilization occurs when unfolded molecules come close enough to combine intermolecularly. Such molecular condensations usually occur during drying, freezing, heating and neutralization of molecular charges of protein solutions. Therefore, the processes associated with irreversible insolubilization can be classified by the patterns of these molecular condensations. Soybean proteins irreversibly insolubilized through neutralization of charges are widely used in the production of tofu in the Orient.

The present paper deals with these changes and their use for food production.

Deteriorative Changes of Soybean Protein During Drying and Their Use in Foods

Irreversible Insolubilization of Soybean Protein During Drying.
Soymilk is an economical high-protein food of high nutritive value produced by grinding soaked whole soybeans with water, heat-

0-8412-0543-4/80/47-123-211$07.25/0

ing the resultant mixture, and then removing the residue to give a stable emulsion. About 65%, 83%, and 73%, respectively, of total solids, protein, and fat contained in whole soybeans are found in the soy milk. Anti-nutritive factors in soybeans, such as trypsin inhibitors, hemagglutinins, etc., are also extracted. Therefore, soy milk must be heated for inactivation of these anti-nutritive factors as well as avoidance of off-flavors. However, soy milk powder produced from a heated soy milk is not easily dispersed into water when reconstituted before drinking. This is a result of insolubilization of the heated protein which occurred during drying. This insolubilization occurs even when drying is carried out at room temperature or by lyophilization. This indicates that the process of evaporation of water during drying is responsible for the insolubilization.

Effect of the heating conditions of soy milk before drying on its redispersibility after drying is not simple. The upper two curves (designated as (a)) in Figure 1 show the effect of temperature and time of heating before drying of soy milk on the amounts of the protein insolubilized during drying. In this figure, 10 ml of heated soy milk in a 250 ml beaker was dried in a 50°C constant temperature room for 16 hours. According to Figure 1, insolubilization during drying of raw soy milk without heating was small; insolubilization was at a maximum after 10 minutes of heating and then decreased gradually with longer heating times at 100° and 120° C.

In order to determine the mechanism of this insolubilization, -SH blocking reagents were added to soy milk heated under the condition which caused maximum insolubilization during drying. The resultant soy milk was dried and the amount of insolubilized protein was measured, as shown in Figure 2. As shown in this figure, the amount of insolubilized protein decreased sharply with the addition of N-ethylmaleimide (NEMI) or sodium-p-chloromercuribenzoate (PCMB) and reached a constant value at around 2×10^{-4} M of either reagent. As the concentrations of free -SH groups of this soy milk were around 4 and 2×10^{-4} M in unheated and heated milk, respectively, this concentration of NEMI or PCMB coincides with the concentration of free -SH groups in the heated soy milk. This may indicate that the free -SH groups present in the heated soy milk protein take part in the insolubilization of the protein during drying.

There are two mechanisms for the molecular polymerization by disulfide bonds. One is the polymerization through an intermolecular disulfide bond formed by oxidation between the two free -SH groups located on different protein molecules. The other mechanism is polymerization through an intermolecular disulfide bond formed by an interchange reaction between free -SH groups and disulfide bonds which are located intermolecularly.

In order to determine whether the disulfide polymerization of heated soy milk protein occurs through the first or second mechanism, it is necessary to measure the disulfide and -SH content of

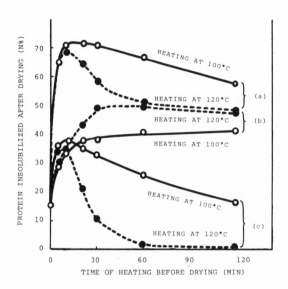

Cereal Chemistry

Figure 1. Effect of heating of soy milk before drying and effect of addition of N-ethylmaleimide (NEMI) to heated soy milk on the insolubilization of protein after drying. The curves are: (a), dried without adding NEMI; (b), dried after adding NEMI; and (C), the values of (a) minus the values of (b). Curve (a) indicates total amount of insolubilized protein; curve (b) indicates the amount of protein insolubilized by mechanisms other than by intermolecular disulfide bond formation; and curve (c) indicates the amount of protein insolubilized through disulfide bond polymerization (3).

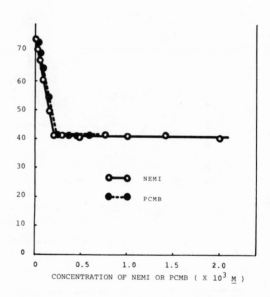

Cereal Chemistry

Figure 2. Effect of the concentration of N-ethylmaleimide (NEMI) or sodium p-chloromercuribenzoate (PCMB) added to heated soy milk (100°C, 20 min) before drying on the insolubilization of the soy milk protein after drying (3).

the protein. The heated soy milk protein was found to contain
only one or two -SH groups per mole (average molecular weight)
whereas there were ten times as many disulfide groups. The
presence of only one or two free -SH groups per mole of soy milk
protein rules out disulfide polymerization via oxidation of -SH
groups because: (1) the probability that the one or two -SH groups
will react intermolecularly is small, (2) even though two -SH
groups might react intermolecularly the molecules will polymerize
only to dimers when each molecule has one -SH group, and to one-
dimensional polymers only if all the molecules contain two -SH
groups. Therefore, large amounts of insolubilized protein through
this mechanism cannot be expected. On the other hand, existence
of large numbers of disulfide bonds in each molecule suggests that
interchain disulfide polymerization of the heated soy milk protein
occurs through an interchange reaction between -SH and disulfide
groups. One or two -SH groups in each molecule could react readily with any of the several accessible disulfide bonds of another
molecule and consequently the interchange reaction between the -SH
and disulfide groups occurs to form a new intermolecular disulfide
bond. By this reaction a new free -SH group appears, which can
take part in another intermolecular reaction with disulfide bonds
to form a new intermolecular disulfide bond. Thus, the interchange reaction can proceed successively, producing new intermolecular disulfide bonds and new -SH groups, as shown in Figure
3. Through this interchange mechanism, the intermolecular disulfide bonds can link at multiple sites on each molecule, resulting in a three-dimensional polymerization and insolubilization of
the molecules.

There is another observation which indicates that polymerization during drying in heated soy milk occurs through the disulfide
bond interchange reaction. Insolubilization of the protein was
increased rather than decreased by the addition of a small amount
of disulfide bond splitting reagents such as mercaptoethanol,
sodium sulfite, etc., as shown in Figure 4. Further addition of
these reagents decreased insolubilization. The explanation appears to be as follows. In the presence of large amounts of disulfide bond-splitting reagents, all the disulfide bonds of the
protein are split and formation of intermolecular disulfide bonds
does not occur. However, in the presence of a small amount of
these reagents, for example 10^{-3} M, some of the disulfide bonds
are split to produce new -SH groups. In this case, some of the
disulfide bonds are not split. Therefore, the increase in -SH
groups act as initiators of the interchange reaction. Thus, insolubilization through polymerization is increased by the disulfide bond interchange reaction with the increased number of -SH
groups. These results agree with those for the -SH and disulfide
bond interchange reactions in plasma albumin (5,6).

Since part of the insolubilization of heated soy milk protein
during drying occurs through the sulfhydryl/disulfide bond interchange reaction, experiments were carried out so as to distinguish

(1) UNFOLDING OF A POLYPEPTIDE CHAIN IN A NATIVE MOLECULE

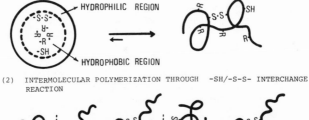

(2) INTERMOLECULAR POLYMERIZATION THROUGH -SH/-S-S- INTERCHANGE
     REACTION

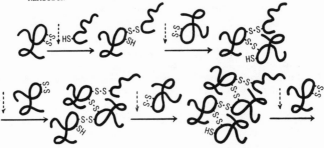

(3) INTERMOLECULAR POLYMERIZATION THROUGH HYDROPHOBIC INTERACTION

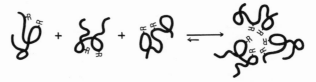

Figure 3.   Schematic diagram of unfolding of native protein molecules and their
intermolecular polymerization (7)

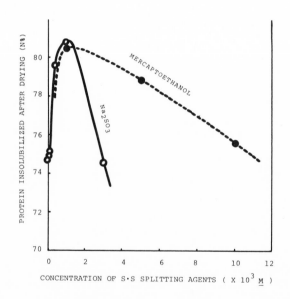

*Figure 4. Effect of added mercaptoethanol and Na₂SO₃ on the insolubilization
of soy milk proteins after drying (2).*

quantitatively between protein insolubilized through the sulf-
hydryl/disulfide bond interchange reaction and protein insolubilized by other methods. The free -SH groups of soy milk heated for
the times indicated in Figure 1 were then blocked by N-ethylmaleimide (NEMI), the protein dried and the amount of insolubilized
protein determined. The results are shown in the middle two
curves designated as (b) in Figure 1. It is quite clear from the
difference between the curves with and without NEMI that some of
the insolubilization of protein occurs through the sulfhydryl/di-
sulfide interchange reaction, which is shown in the lower two
curves designated (c) in Figure 1. The decrease of extent of insolubilization through disulfide bond polymerization resulting
from prolonged heating, (Fig. 1(c)), indicates that loss of some
of the free -SH groups of the protein occurred during the heating.
This -SH loss was more rapid at 120°C than at 100°C and is attributed to oxidation by O_2 present in soy milk. Free -SH groups
react rapidly in the presence of oxygen at high temperature, forming (probably) sulfenic acid (-SOH), sulfinic acid (-SO_2H), and
sulfonic acid (SO_3H) groups. Oxidative reagents, such as hydrogen
peroxide, also remove the free -SH groups of soy milk protein.
      Next, in order to determine whether the insolubilization
brought about by means other than polymerization through disulfide
bonds is due to hydrophobic bonds, the solubility behavior of insolubilized protein in which the -SH groups of soy milk were
blocked was tested after drying with sodium dodecylsulfate (SDS),
a hydrophobic bond disrupting agent. Almost all the protein insolubilized during drying of the -SH blocked soy milk were solubilized by 0.5% SDS at neutral pH, indicating some insolubilization occurs by means other than polymerization by sulfhydryl/di-
sulfide bond interchange. This insolubilization may be due to
intermolecular polymerization through hydrophobic interactions.
      Thus, it is concluded that insolubilization of soy milk pro-
tein during drying occurs both through intermolecular disulfide
bonds formed by interchange between the -SH and disulfide groups
of the molecules and through intermolecular hydrophobic interaction. When soy milk was not heated before drying and, therefore
the proteins in soy milk were in a native state, the insolubilized
protein after drying was around 16%, regardless of the presence
or absence of NEMI. This may be a result of few or no disulfide
bonds on the surface of the native protein to interchange with the
-SH groups. When soy milk is heated, however, the native three-
dimensional structure of the molecules are disrupted and as a result the free -SH groups, many disulfide bonds, and most of the
hydrophobic groups, formerly buried inside the molecules, are exposed (Fig. 3). When the exposed residues are brought into close
proximity as a result of drying both disulfide bonds and hydrophobic bonds, formed intermolecularly by mechanisms described a-
bove, contribute to the insolubilization of soy milk. Longer
times and higher temperatures during heating of soy milk before
drying increased the number of exposed hydrophobic groups, in-

creasing the amount of protein insolubilized by intermolecular hydrophobic interaction (middle two curves of (b) in Fig. 1). Thus, the complicated phenomena observed for insolubilization of the protein of heated soy milk may be explained by these mechanisms.

There is another phenomenon, regarded as a deteriorative change in the protein of soy milk, caused also by the evaporation of water. This is a film formation on the surface of soy milk, which occurs when heated soy milk is kept open to the air. This phenomenon is observed not only in heated soy milk but also in heated cow's milk. Film formation of soy milk occurs only when the soy milk is heated above 60°C and there is evaporation of water from the surface of the soy milk. The mechanism of protein insolubilization is basically the same as that of soy milk powder produced from heated soy milk (4). When water is removed from the surface of heated soy milk by evaporation, the molecular concentration of protein near the surface increases locally and the exposed reactive groups of the denatured molecules come close enough to interact intermolecularly both by hydrophobic interactions and through the sulfhydryl/disulfide interchange reaction to form a polymerization (film) on the surface. The upper side of the film contains more hydrophobic amino acids because of orientation of the hydrophobic portions of the unfolded molecules to the atmosphere rather than into the aqueous solution.

Use of Deteriorative Changes of Protein During Evaporation for Food Production. In Japan, there is a traditional product called "yuba" manufactured by irreversible insolubilization of soy milk protein during evaporation. The film formation of heated soy milk described above is utilized for production of yuba. Yuba production was studied in detail by Okamoto et al. (8,9). In the making of yuba, soy milk is put into an open, shallow pan and heated above 80 °C. The film, formed on the surface by heating and evaporation of water, is skimmed from the surface repetitively with a fine stick and dried by warm air. More than 80% of the soy milk solids can be recovered from soy milk as yuba. Samples of yuba are shown in Figure 5. Yuba is a very nutritious protein food composed of 8.7% water, 52.3% protein, 24.1% fat, 11.9% carbohydrate, and 3.0% ash. Yuba, with a meat-like texture, is used as an ingredient in various dishes after seasoning.

Deteriorative Changes of Soybean Protein After Freezing and Their Use for Foods

Irreversible Insolubilization of Soybean Protein After Freezing. It is well known that deteriorative changes occur in proteins during frozen storage. Hashizume et al. (10,11,12) have investigated the insolubilization of soybean protein after freezing. Comparison of these results with those of insolubilization of heated soy milk protein during drying indicate that protein insol-

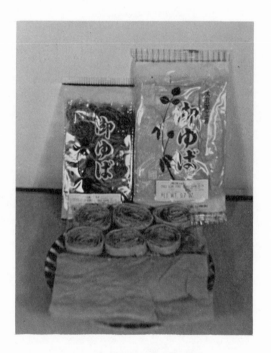

Figure 5. Samples of "yuba"

ubilization after freezing and during drying occur essentially by the same mechanisms. Both are insolubilized through sulfhydryl/disulfide interchange reactions and hydrophobic interractions, which occur when protein molecules are brought close together through concentration. In the evaporation the molecules are brought enough to react intermolecularly by removal of water, whereas in freezing they come close enough to react by removal of water by formation of ice crystals. When a protein solution is frozen, the protein molecules are concentrated into the unfrozen water solution which exists among the ice crystals. The amount of this unfrozen water depends upon the rate of freezing and temperature used, the temperature of frozen storage and the kind of solution in which the proteins were dissolved. The lower the temperature, the lower the amount of unfrozen water. At a very low temperature, such as -30°C, the amount of unfrozen water is very small and therefore, insolubilization will not occur by the above mechanisms if the protein solution is frozen rapidly. At -3°to -5°C, however, most frozen foods contain about 10-20% of unfrozen water among the ice crystals in which the protein molecules are concentrated. In such concentrated solutions, the sulfhydryl/disulfide interchange reaction and hydrophobic interaction described above can occur readily, resulting in protein insolubilization.

Insolubilization of heated soybean protein after freezing occurs for the same reasons as insolubilization of heated soybean protein during drying. Figure 6 shows the effect of time of frozen storage on the insolubilization of heated soybean protein solution following freezing and the solubility of the insolubilized protein in urea and/or mercaptoethanol. As shown by curve (a) in Figure 6, the heated soybean protein was insolubilized rapidly during frozen storage. Most of the insolubilized protein could be solubilized by urea alone, as long as the time of frozen storage is short, but the protein became more insoluble in urea as the time of frozen storage increased (curve (b)). All the insoluble protein could not be solubilized, until mercaptoethanol is added to urea (curve (c)). This indicates that insolubilization of heated proteins during frozen storage occurs mainly by hydrophobic bonds in the initial stage of the storage but disulfide bonds are gradually formed on longer storage. As a result, the protein was not solubilized by urea only.

The disulfide bonds formed during frozen storage are probably formed through sulfhydryl/disulfide interchange reactions, just as in heated soy milk powder during drying, since insolubilization during frozen storage was also accelerated by addition of small amounts of mercaptoethanol (Fig. 7).

In order to determine whether formation of these hydrophobic and disulfide bonds was caused by concentrating the protein molecules into the liquid phase among the ice crystals, heated soybean protein solution was concentrated to about 60% water content at room temperature using carbowax (polyethylene glycol 6000) and

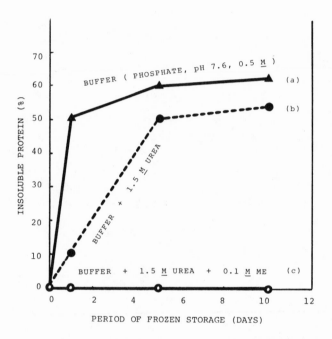

Agricultural and Biological Chemistry

Figure 6.   The insolubilization of soybean protein during frozen storage at −5°C and their solubility behavior in urea and mercaptoethanol (ME) (10).

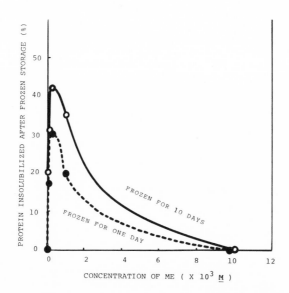

Figure 7. Increase in the insolubilization of soybean protein during frozen stor-age at −5°C by the addition of small amounts of mercaptoethanol (ME), indi-cating the promotion of a sulfhydryl/disulfide interchange reaction by a disulfide bond splitting agent (10)

then stored at +5°C without being frozen. Measurements of amount
of insolubilized protein and its solubility in urea and mercaptoethanol are shown in Figure 8. The results show that protein insolubilization occurred in the sample stored in a concentrated
state without being frozen just as it did in the sample stored in
a frozen state without prior concentration (Fig. 6). Similar solubility curves in urea and mercaptoethanol were obtained for these
two samples (Figs. 6 and 8), indicating that insolubilization resulted from concentrating the solutions (by evaporation or by
freezing) permitting sulfhydryl/disulfide interchange and hydrophobic reactions to occur.

     As described above, insolubilization does not occur generally
in native proteins. However, if a native protein has -SH and disulfide groups on its surface, insolubilization may occur during
frozen storage even in native protein molecules. In this insolubilization, contribution of the hydrophobic bonds is less than in
the denatured protein because most of the hydrophobic residues are
buried inside the native protein molecule. Therefore, the contribution of disulfide bonds to protein insolubilization during
frozen storage can be assessed more clearly in native proteins
than in heated ones. For example, 11S globulin (glycinine), one
of the major components of soybean globulins (Table 1), contains
-SH and disulfide groups on the surface of the native molecule.
When a solution of native 11S globulin was stored in the frozen
state at -5°C, precipitation occurred and several polymerized
molecules were present even in the supernatant as shown in Figure
9(b) of the disc-gel electrophoretic patterns. Addition of di-
sulfide-splitting agents, such as mercaptoethanol, to this solution stored in the frozen state, however, solubilized the precipitates and the soluble polymers were depolymerized as shown in
Figure 9(c). Moreover, when -SH blocking agents, such as NEMI,
were added to native 11S globulin solution before freezing insolubilization did not occur on frozen storage. These observations
indicate that insolubilization of native 11S globulin during
frozen storage occurred primarily through disulfide bond formation
and hydrophobic bonds were not primarily responsible for this insolubilization.

     When the 11S globulin solution was stored in a concentrated
state without being frozen, polymerization of the molecules occurred; they had the same solubility behavior as solutions stored in
a frozen state without being concentrated as shown in Figure 9(d)
and (e). Therefore, the disulfide bonds formed during frozen
storage of native 11S globulin, as well as in heated soybean protein, were caused by the concentrating of the protein molecules
into the liquid phase among the ice crystals.

     A special sponge-like texture produced as a result of insol-
ubilization of unfolded protein molecules through both hydrophobic
interactions and sulfhydryl/disulfide interchange reactions, caused by concentrating the proteins into the liquid phase present a-
mong the ice crystals during frozen storage, is shown in Figure 10.

Table 1. Major Protein Components of Soybean*

Protein Components		Protein Contents		Molecular Features	
By ultra-centrifuge	By immu-nology	By immu-nology	By ultra-centrifuge	M.W.	Half cystine (No. per mol)
2S Globulin	α-Con-glycinine	13.8%	15.0%	32,600	6
7S Globulin	β-Con-glycinine	27.9%	34.0%	180,000	4
	γ-Con-glycinine	3.0%		104,000	--
11S Globulin	Glycinine	40.0%	41.9%	360,000	48
15S Globulin	--	--	9.1%	--	--

*Taken from Table I of Fukushima (7).

Journal of Japan Soy Sauce Research Institute

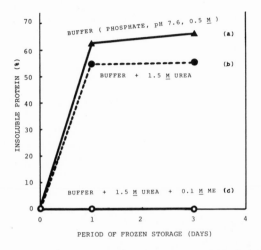

Agricultural and Biological Chemistry

Figure 8. The insolubilization of soybean protein during storage at +5°C in a concentrated state and its solubility behavior in urea and mercaptoethanol (ME) (10).

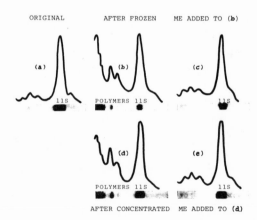

Agricultural and Biological Chemistry

Figure 9. Disc-gel electrophoretic patterns of 11S soybean globulin stored in a frozen or concentrated state. (a), original solution; (b), after 2 days storage in a frozen state at −5°C; (c), after the addition of 0.01M mercaptoethanol (ME) to solution (b); (d), after 2 days of storage in a concentrated state (unfrozen); and (e), after the addition of 0.01M mercaptoethanol to solution (d) (10).

Use of Deteriorative Changes in Protein During Frozen Storage
in Food Production. There is a very unique product made by use of
the changes described above during frozen storage. This is a soybean protein product called "kori-tofu" which was originally developed in ancient Japan in the regions with severely cold
winters. The first step of kori-tofu making is the production of
soy milk curd from soy milk, using calcium salts as a coagulant,
just as the first step of cheese making is that of milk curd from
cow's milk using rennet as a coagulant. The initial soy milk
curd, called "tofu", possesses a fragile and gelatinous texture as
described later. The second step of kori-tofu making is frozen
storage of the tofu curd. The tofu curd is frozen at -10°C rapidly and then kept at -1° to -3°C for 2 to 3 weeks. During this
frozen storage, intermolecular interaction of the protein occurs
in the liquid phase which surrounds each crystal through the mechanisms described above. As a result, the texture of the soy
milk curd after thawing has changed dramatically from a fragile
and gelatinous texture to a strong and sponge-like texture with a
great many holes where the ice crystals existed. The final step
of kori-tofu making is a drying process. After thawing, the drying can be carried out very easily by first squeezing out most of
the water inside the curd and then blowing a warm air current on
the material. The final product is usually 20 gram square pieces
as shown in Figure 11. It is a very nutritious product which contains (typically) 53.5% protein, 26.5% oil, 7.0% carbohydrate,
2.5% ash, and 10.5% water. Before preparation for eating, kori-
tofu is reconstituted by soaking in hot water. The rehydrated
kori-tofu can imbibe a large amount of seasoning solution and is
usually used as an ingredient in various dishes after cooking with
seasonings. A meat-like chewiness and flavor can be given to the
reconstituted kori-tofu, depending upon the method of cooking.
Kori-tofu is mass produced in modern factories where about 30,000
metric tons of soybeans are used for its production annually in
Japan.

Reversible and Irreversible Insolubilization of Soybean Protein
and Their Use for Foods

It is very important in food processing whether soybean protein is reversibly or irreversibly insolubilized, since irreversible insolubilization generally results in deterioration of the
physical properties of the protein. Irreversible insolubilization
occurs when unfolded molecules are brought close enough, through
water evaporation, freezing of water or the neutralization of molecular charges, to form intermolecular bonds. In Figure 12, reversible and irreversible insolubilizations are classified schematically according to the patterns of the condensation of the
molecules.

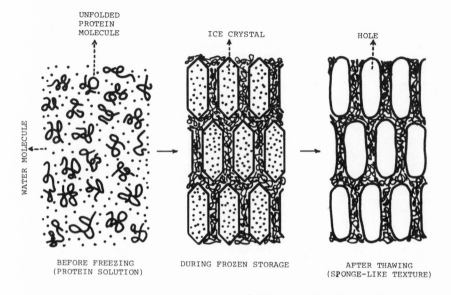

UNFOLDED
PROTEIN
MOLECULE

ICE CRYSTAL

HOLE

WATER MOLECULE

BEFORE FREEZING
(PROTEIN SOLUTION)

DURING FROZEN STORAGE

AFTER THAWING
(SPONGE-LIKE TEXTURE)

Japanese Society of Food Science and Technology

Figure 10.   Schematic diagram of the insolubilization of soybean protein during frozen storage (11)

Figure 11.   Samples of "kori-tofu"

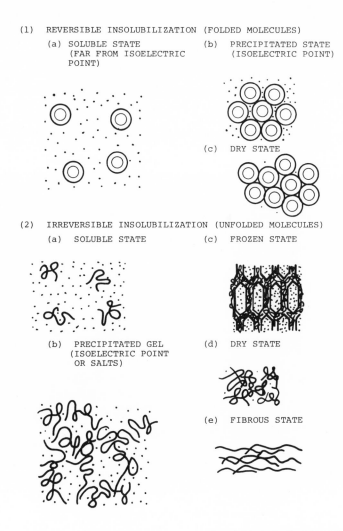

(1) REVERSIBLE INSOLUBILIZATION (FOLDED MOLECULES)

    (a) SOLUBLE STATE   (b) PRECIPITATED STATE
        (FAR FROM ISOELECTRIC   (ISOELECTRIC POINT)
        POINT)

                                        (c) DRY STATE

(2) IRREVERSIBLE INSOLUBILIZATION (UNFOLDED MOLECULES)

    (a) SOLUBLE STATE   (c) FROZEN STATE

    (b) PRECIPITATED GEL   (d) DRY STATE
        (ISOELECTRIC POINT
        OR SALTS)

                                        (e) FIBROUS STATE

Journal of Japan Soy Sauce Research Institute

Figure 12. Schematic diagram for the mechanisms of reversible and irreversible insolubilization of soybean protein (7)

Usually, reversible insolubilization occurs when the protein molecules are in a native state as shown in Figure 12-(1). The surface of native protein molecules contains primarily the hydrophilic amino acid residues and, even though molecules may contact each other during isoelectric precipitation, through concentration by evaporation of water and by freezing, irreversible intermolecular bonds are not generally formed among the molecules. Even in the native state however, irreversible insolubilization through a sulfhydryl/disulfide interchange reaction may occur when free -SH and disulfide bonds are located at the surface of the molecules. This has already been described above for the insolubilization of native 11S globulin during frozen storage.

Free -SH groups are also very sensitive to oxidation even by air. Native soybean protein in solution became less soluble when frozen immediately after preparation than it did when frozen after storage for 2 days (Fig. 13). The same behavior was found for heated soybean protein frozen immediately or after two days. These results tend to indicate that the one or two -SH groups in soybean protein become oxidized after storage for two days in the unfrozen state. Thus, intermolecular disulfide bond formation could not occur as described in Figure 3.

Reversibly insolubilized soybean protein products possess various functional properties, such as binding, emulsification effect, etc. These functionalities may appear when the native protein molecules are unfolded during heating in food processing. Therefore these products, such as soybean protein isolate, are useful as binders or emulsifiers for sausage, hams, etc.

On the other hand, irreversible insolubilization occurs among unfolded protein molecules. In unfolded soybean protein molecules, the -SH, disulfide and hydrophobic amino acid side chains of the molecules are exposed, but the molecules remain soluble when the concentration is not too high, as shown in Figure 12(2a). A typical example of this type of product is soy milk. When the unfolded soybean protein molecules are concentrated so that contact among them is enhanced, however, irreversible insolubilization occurs through both sulfhydryl/disulfide interchange and hydrophobic interactions. As described above, molecules may be brought together by concentration of the molecules through removal of water by evaporation and through removal of water by freezing. Other methods of bringing the molecules together are through neutralization of charges by adding salt or acidifying agents, and by extension and orientation of proteins.

A typical example of charge neutralization in food production is the manufacture of tofu, a soybean protein food consumed in large amounts in Japan. When calcium sulfate is added to heated soy milk, the soy milk is coagulated. This is due to decrease of the negative charge on the protein as a result of binding of Ca^{2+} to the negatively charged acidic amino acid residues of the protein molecules. Therefore the unfolded molecules can aggregate, owing to the decrease of electrostatic repulsion, and then form

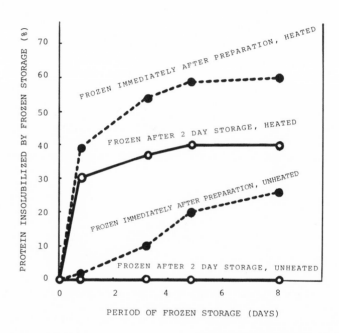

Japanese Society of Food Science and Technology

Figure 13. Comparison of rates of insolubilization during frozen storage between soybean protein solutions frozen immediately after preparation (heated and unheated) and frozen after 2 days of storage (heated and unheated). The heated samples were held at 100°C for 5 min prior to freezing (11).

an irreversible coagulate. Instead of calcium salts, glucono-δ-
lactone is often used. The glucono-δ-lactone is hydrolyzed to
gluconic acid during heating and acts as an acidifying agent. In
this case, the negative charge on the protein is decreased by
protonation of the -COO⁻ of the acidic amino acid residues.

Tofu is a white gelatinous curd with a unique texture in
which large amounts of water are held (Fig. 14). The texture is
soft, smooth, and elastic. Typical percentages of water, protein,
oil, carbohydrate and ash in tofu are 88.0%, 6.0%, 3.5%, 1.9% and
0.6%, respectively. In Japan, 270,000 metric tons of whole soybeans and 65,000 metric tons of defatted soybean meal are used in
making tofu and its derivatives.

An example where extension and orientation of protein molecules is used to bring them together for interaction is in artificial meat products, including textured protein products. In
such products, disulfide bonds, hydrophobic bonds and hydrogen
bonds are formed among the proteins extended as fibers as shown
in Figure 12(2e).

Irreversible insolubilization of proteins may occur mainly
through formation of both intermolecular disulfide and hydrophobic
bonds. The product can be quite different depending on the relative contribution of these two types of bonds. The hydrophobic
bonds are formed among the hydrophobic amino acid side chains contributed by valine, leucine, isoleucine, phenylalanine, etc.
These side chains share a common lack of affinity for water and
are pushed together out of the network of water molecules in order
that water may preserve its structure. Each hydrophobic bond is a
weak bond (1-2 kcal/mole), but they may make a significant contribution to stabilization of the polymerized state if there are
enough exposed hydrophobic residues among the molecules. In
contrast, disulfide bonds are covalent and strong (80-100 kcal/
mole). Therefore, the amount of intermolecular disulfide bond
formation will have a major influence on the physical properties
of the insolubilized proteins. For instance, there is a marked
difference between the physical properties of tofu gel made from
7S and 11S globulins. 7S globulin (β-conglycinine) does not contain free -SH groups and only two disulfide bonds per molecule,
whereas 11S globulin has a number of free -SH groups and a large
number of disulfide bonds (Table 1). Therefore, the tofu gel made
from 7S globulin is mostly stabilized by hydrophobic bonds, while
the tofu gel made from 11S globulin is stabilized by both disulfide bonds formed through the sulfhydryl/disulfide interchange
reaction and hydrophobic bonds. This is the reason why 7S tofu
gel is soft and less elastic, while 11S tofu gel is much more
elastic (10,13). The same differences can be seen between the
physical properties of yuba produced from 7S and 11S globulins.
Yuba film made from 11S protein is much stronger than when made
from 7S protein (14).

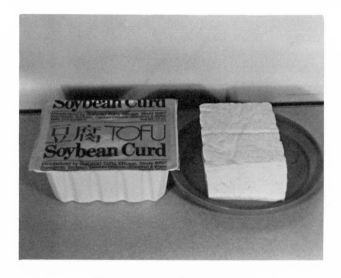

Figure 14. Samples of "tofu"

Effect of Deteriorative Changes of Soybean Protein During Heating
on Enzyme Digestibility

Enzyme Digestibility and Yield of Soy Sauce. There are
various kinds of traditional soybean protein foods in the Orient.
In addition to soy milk, tofu, kori-tofu and yuba described so
far, there are fermented soy sauce, miso, natto, sufu and temphe.
Soy sauce was introduced into Japan during the 7th century by
Buddhist priests and has been developed into the present-day
Japanese type of soy sauce, characterized by an excellent aroma
and flavor, through centuries of artistry. Recently, fermented
soy sauce has become popular with Western people.
     Manufacture of fermented soy sauce is composed of three pro-
cesses, the koji making process, the brine fermentation process,
and the refining process. For koji production, Aspergillus
species are inoculated onto the cooked solid mixture of soybeans
and wheat and cultured for 40 to 45 hours under circulating air of
constant temperature and humidity. The cultured solid mash,
called koji, is then mixed with a brine (NaCl) solution of 14 to
15 percent by weight. During this brine fermentation, the protein
in the soybeans and wheat is hydrolyzed by proteases from the
Aspergillus species. This is in contrast to a chemical soy sauce
made by hydrolysis of proteins with HCl. Therefore, digestibility
of the proteins by enzymes is one of the most important factors in
the making of fermented soy sauce because it is closely related to
the yield of soy sauce.
     The digestibility of soybean and wheat proteins by the en-
zymes is markedly influenced by the conditions of heat treatment
of the soybeans. Native soybean protein is quite resistant to
proteolysis because of its compact conformation. The rate of
proteolysis is dependent on the degree of unfolding of the substrate protein molecules as shown in Figure 15. Accordingly, when
soybean protein is used as substrate for proteases, the protein
molecules must be unfolded by some treatment, such as heating.
However, heat treatment of the protein may decrease the rate of
proteolysis. Extended heating of soybean protein decreases the
rate of proteolysis as shown in Figure 16. Therefore, denaturation of the protein leads to better proteolysis but too much heat
treatment decreases the rate of proteolysis by causing other
changes in the protein.
     Factors which affect the rate and extent of enzymatic hydro-
lysis of proteins include: (1) the substrate specificity of the
enzymes, (2) modification of the amino acid side chains of the
substrate proteins, and (3) the three-dimensional structure of the
substrate proteins. It is essential that the active center of the
enzymes be able to bind with specific amino acid residues of the
substrate protein. The native soybean protein molecules are completely folded and therefore the specific amino acid residues required by the enzymes may not be available. This is why native

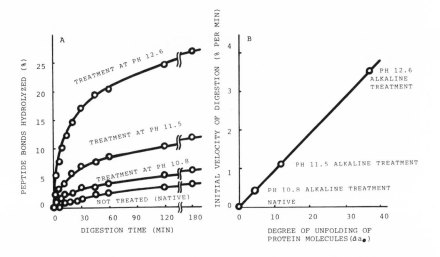

Figure 15. Relationship between the degree of unfolding of 11S globulin molecules and their susceptibility to proteolysis. In Figure 15B, Δa_0 is calculated as $\left(a_0^{Sample} - a_0^{Native}\right)/\left(a_0^{Urea-denatured} - a_0^{Native}\right)$ in the Moffitt–Young equation for optical rotatory dispersion. The samples were treated at 20°C for 90 min at the indicated pH and then neutralized (15).

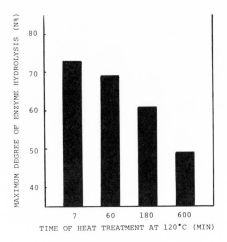

Figure 16. The effect of heat treatment of soybean protein on the maximum extent of enzymatic hydrolysis by proteases of Aspergillus species (4)

soybean protein cannot be hydrolyzed by enzymes readily. In contrast, when the protein is unfolded by heat treatment enzymatic
hydrolysis will proceed rapidly because the enzyme-specific amino
acid residues of the substrate are available as shown in Figure
17.

However, it is known that some amino acid residues of proteins are modified during heating through reaction with other compounds or through cross-linking. For instance, α- and ϵ-amino
groups may be modified by reaction with aldehyde compounds such as
glucose, while lysine, serine, cystine, threonine, arginine, histidine, tryptophan, aspartic acid and glutamic acid may be modified to lysinoalanine or other compounds through β-elimination and
cross-linking during heat treatment of proteins (14-19).

The alkaline proteases of Aspergillus species used in soy
sauce manufacture are specific for tyrosine, phenylalanine, leucine, lysine, and arginine residues in proteins. It has been
shown that lysine, arginine and cystine of the soybean proteins
are partly destroyed or modified during heat treatment of defatted
soybean flour in the presence of water (Table 2). Since some of
these amino acids are essential for maximum hydrolysis by the enzymes of Aspergillus species, their destruction or modification
will result in a decrease in the degree of maximum hydrolysis by
enzyme. This is one of the reasons why maximum hydrolysis of the
protein was decreased by prolonged heating (Fig. 16). Also, during prolonged heating new intermolecular or intramolecular interactions among the hydrophobic residues of the unfolded protein
will also result in a decrease of enzymatic hydrolysis.

With due consideration of the effect of heating on digestibility of soybean protein, various investigations were carried out
using high-temperature - short-time treatment for denaturation of
the soybean protein for use in making soy sauce. A high temperature treatment achieved maximum unfolding of the soybean protein.
A very short time treatment minimized the other deteriorative
changes. Therefore the yield of soy sauce, based on weight of
protein of the starting soybean, has increased from 65% of 20
years ago to almost 90% at the present.

Enzyme Digestibility and Nutritive Value of Protein. Decreased digestibility of soybean protein with an increase of time
of heat treatment is also observed for trypsin and pepsin as well
as for the enzymes from Aspergillus species. This decrease in
trypsin and pepsin digestibility gives decreased nutritive values,
just as the decrease in hydrolysis by the enzymes from Aspergillus
species gave a decrease in yield of soy sauce. The influence on
digestibility of the destruction of amino acid side chain residues
during heating will be larger for trypsin than for other proteases
because the amino acids, lysine and arginene, specific for tryptic
hydrolysis are more sensitive to destruction during heating(Table
2). The action of trypsin on unheated soybean protein preparations is particularly low in comparison with other enzymes. This

(1) ENZYMATIC HYDROLYSIS OF FOLDED PROTEIN MOLECULE (NATIVE PROTEIN)

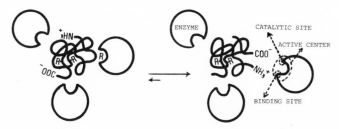

(2) ENZYMATIC HYDROLYSIS OF UNFOLDED PROTEIN MOLECULE (DENATURED PROTEIN)

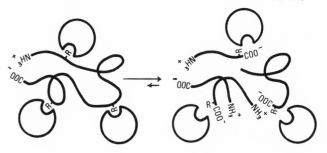

Journal of Japan Soy Sauce Research Institute

Figure 17. Schematic explanation for enzymatic hydrolysis of denatured proteins
(7)

Table 2.   The Destruction of Some Amino Acids During the
Heating of Defatted Soybean Flour Protein*

Amino acids	No heat-treatment	Hours at 126°C				Hours at 115°C			
		0.5	1	2	4	0.5	1	2	4
		(Amino acid residue %/100 gr. protein)							
Gly	4.1	4.1	4.1	4.2	4.0	4.2	4.1	4.0	4.1
Ala	4.4	4.4	4.5	4.6	4.5	4.4	4.5	4.3	4.2
Val	5.4	5.5	5.4	5.5	5.5	5.7	5.5	5.5	5.1
Ile	5.2	5.1	5.2	5.0	5.0	5.1	5.2	5.1	4.9
Leu	8.4	8.5	8.4	8.5	8.5	8.4	8.5	8.4	8.3
Asp	12.2	12.0	12.2	12.0	12.2	12.1	12.1	12.2	12.2
Glu	19.7	19.2	19.5	19.4	19.8	19.5	19.5	19.6	19.5
Lys	6.3	6.0	5.9	5.8	5.4	5.6	5.6	5.1	4.0
Arg	7.6	7.5	6.8	6.3	6.1	6.2	6.3	5.9	4.5
His	2.4	2.6	2.3	2.3	2.3	2.5	2.3	2.5	2.4
Phe	5.1	4.9	5.3	5.1	5.1	5.2	5.2	5.0	5.0
Tyr	3.3	3.2	3.1	3.2	3.3	3.2	3.3	3.2	3.1
Pro	5.5	5.4	5.5	5.4	5.6	5.3	5.4	5.5	5.5
Trp	1.1	1.1	1.1	1.1	1.1	1.0	1.1	1.0	0.9
Met	0.98	1.0	1.0	1.0	1.1	1.0	1.0	1.0	1.0
Half Cys	1.3	1.3	1.2	1.1	0.9	1.2	1.0	1.0	0.8
Ser	6.4	6.3	6.2	6.1	6.2	6.0	6.2	6.0	5.7
Thr	4.6	4.5	4.6	4.5	4.4	4.5	4.6	4.6	4.5

*Taken from Table I of Taira et al. (20).

Agricultural and Biological Chemistry

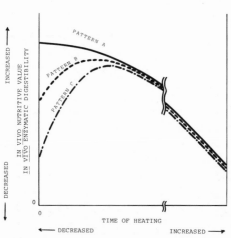

Japanese Society of Miso Science and Technology

Figure 18.   Schematic representation of effect of heat treatment on soybean protein and its hydrolysis patterns by various enzymes. Pattern A, pepsin and other acid proteinases; pattern B, the proteinases having an optimum near neutrality, such as papain, bacteria neutral proteinase, Aspergillus *alkaline proteinase,* Aspergillus *neutral proteinase; and pattern C, trypsin and* in vivo *nutritional values (21).*

is due to trypsin inhibitors which are present in soybeans.
Therefore, increase in tryptic digestibility by heating is attributed to inactivation of trypsin inhibitors as well as unfolding of
the native protein molecules.

The digestion of heated or unheated soybean proteins by various enzymes is schematically compared with the nutritive values in
Figure 18. Pattern A is typical of pepsin where, because of low
pH of the reaction, the protein does not have to be denatured pri-
or to addition to the reaction. Pattern B is typical of enzymes
such as papain, bacterial neutral protease etc. where prior denaturation of the substrate protein is required but there are no
inhibitors of the enzyme present. Pattern C is typical of trypsin
where prior heat treatment of the substrate protein is required to
destroy inhibitors of trypsin as well as to denature the protein
for digestion. The decrease in digestibility with prolonged heating in all three cases is due to modification of the substrate
protein as described above.

Conclusion

Deterioration of the physical properties of proteins during
food processing or food storage can be ascribed primarily to an
irreversible insolubilization of proteins. However, a deteriorative change for one purpose can be a favorable one for another
purpose. In Japan, for instance, the irreversible insolubilization of soybean proteins has been utilized effectively for production of soybean protein foods, such as tofu, kori-tofu, and yuba.

Generally, irreversible insolubilization occurs when unfolded
protein molecules are brought close enough together to combine
intermolecularly. This molecular condensation usually occurs as a
result of evaporation of water, freezing of water, and neutralization of molecular charges which results in intermolecular polymerization among the unfolded molecules. The bonds responsible for
the intermolecular polymerization are both the disulfide bonds
formed by sulfhydryl/disulfide interchange reaction and interaction among the hydrophobic amino acid residues located in the unfolded polypeptide chains of the molecules.

During heating of soybean protein, deteriorative changes may
occur which decrease enzymatic digestibility. These changes are
the result of both the modification of the enzyme-specific amino
acid residues of soybean proteins and hydrophobic bonds formed a-
mong the exposed hydrophobic amino acid residues during prolonged
heating.

Literature Cited

1. Fukushima, D. Proceedings of Western Hemisphere Nutrition
       Congress II 1968, American Medical Association, 1969,
       p. 1.
2. Fukushima, D.; Van Buren, J. P. Cereal Chem., 1970, 47, 571.

3. Fukushima, D.; Van Buren, J. P. Cereal Chem., 1970, 47, 687.
4. Fukushima, D. Cereal Chem., 1969, 46, 405.
5. Huggins, C.; Tapley, D. F.; Jensen, E. V. Nature, 1951, 167,
       592.
6. Hospelhorn, V. D.; Cross, B.; Jensen, E. V. J. Am. Chem.
       Soc., 1954, 76, 2827.
7. Fukushima, D. J. Japan Soy Sauce Res. Inst., 1977, 3, 22.
8. Okamoto, S. J. Japanese Soc. Food Sci. Technol., 1977, 24,
       40.
9. Okamoto, S. Cereal Foods World, 1978, 23, 256.
10. Hashizume, K.; Kakiuchi, K.; Koyama, E.; Watanabe, T. Agr.
       Biol. Chem., 1971, 35, 449.
11. Hashizume, K.; Kosaka, K.; Koyama, E.; Watanabe, T. J.
       Japanese Soc. Food Sci. Technol., 1974, 21, 136.
12. Hashizume, K. Chemistry and Biology (Japan), 1977, 15, 301.
13. Saio, K.; Kamiya, M.; Watanabe, T. Agr. Biol. Chem., 1969,
       33, 1301.
14. Shirai, M.; Watanabe, K.; Okamoto, S. J. Japanese Soc. Food
       Sci. Technol., 1974, 21, 324.
15. Fukushima, D. Cereal Chem., 1968, 45, 203.
16. Hurrell, R. F.; Carpenter, K. J. Abst., ACS, 1976, 172, 42.
17. Finley, J. W.; Friedman, M. Abst., ACS, 1976, 172, 46.
18. Sternberg, M.; Kim, C. Y. Abst., ACS, 1976, 172, 39.
19. Ziegler, K.; Schmitz, I.; Zahn, H. Abst., ACS, 1976, 172,
       101.
20. Taira, Ha; Taira, Hi; Sugimura, K.; Sakurai, Y. Agr. Biol.
       Chem., 1965, 29, 1074.
21. Fukushima, D. Miso Sci. Technol. Japan, 1973, No. 232, p. 2.

RECEIVED October 18, 1979.

Suicide Enzyme Inactivators

BRIAN W. METCALF

Merrell Research Center, Merrell-National Laboratories, Cincinnati, OH 45215

A new and elegant approach to specific irreversible enzyme
inactivation is the use of inhibitors possessing latent reactive
functionalities which are unmasked at the enzyme's active site as
a result of the normal catalytic turnover. Such an inhibitory
process is described by the following equation:

$$E + I \underset{K_I}{\overset{\longrightarrow}{\rightleftharpoons}} E.I \quad \xrightarrow{k_{cat}} E.I' \quad \xrightarrow{k \text{ inact}} E\text{-}I'$$

As these inhibitors owe their activity to the k_{cat} term (i.e.,
the enzyme's usual mode of action) they have been designated
"k_{cat} inhibitors" by Rando (1), while Abeles and Maycock (2) have
used the term "Suicide Enzyme Inactivators" because the enzyme,
in accepting such a "booby-trapped" substrate commits suicide by
its own mechanism of action.

An early example of the concept was described by Wood and
Ingraham who reported that the product of oxidation of phenol or
pyrocatechol by tyrosinase inactivates that enzyme irreversibly(3).
It was speculated that the quinonoid products of oxidation react
in Michael fashion with nucleophilic residues on the enzyme,
leading to covalent binding.

What is now considered as the classical example of this concept was discovered by Endo et al. (4) who described in 1970 the
irreversible inhibition of β-hydroxydecanoylthioester dehydrase
by the propargylic thioester I (Fig. 1). I, being an analogue of
the corresponding *cis* olefin, which is a natural substrate, undergoes proton abstraction by the enzyme to generate a propargylic
anion, which on reprotonation affords the conjugated allene. This
allene, being a Michael acceptor and hence an active alkylating
agent, is able to react with a nucleophilic histidine residue in
the active site of the enzyme leading to covalent bond formation
and irreversible inactivation.

During the last nine years, a number of workers have
attempted to generalize this concept of enzyme inactivation to the
inhibition of enzymes other than β-hydroxydecanoylthioester dehydrase. The reader is referred to excellent reviews by Rando (1),

0-8412-0543-4/80/47-123-241$05.00/0

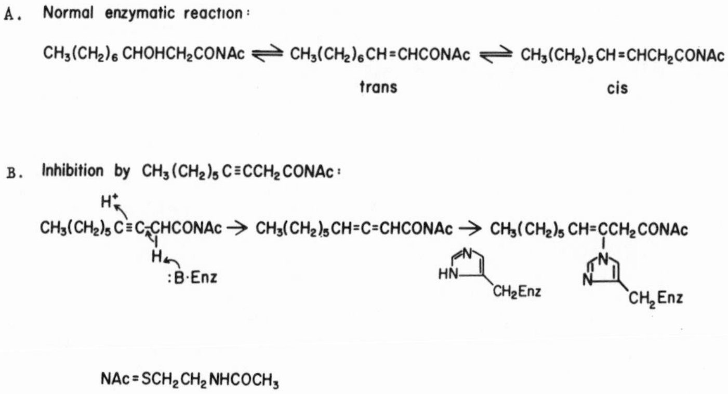

A. Normal enzymatic reaction :

$CH_3(CH_2)_6 CHOHCH_2CONAc \rightleftharpoons CH_3(CH_2)_6CH=CHCONAc \rightleftharpoons CH_3(CH_2)_5CH=CHCH_2CONAc$

trans   cis

B. Inhibition by $CH_3(CH_2)_5 C\equiv CCH_2 CONAc$:

$CH_3(CH_2)_5 C\equiv C-CHCONAc \rightarrow CH_3(CH_2)_5CH=C=CHCONAc \rightarrow CH_3(CH_2)_5 CH=CCH_2CONAc$

:B·Enz

NAc = $SCH_2CH_2 NHCOCH_3$

Figure 1.   Inhibition of β-hydroxydecanoylthioester dehydrase

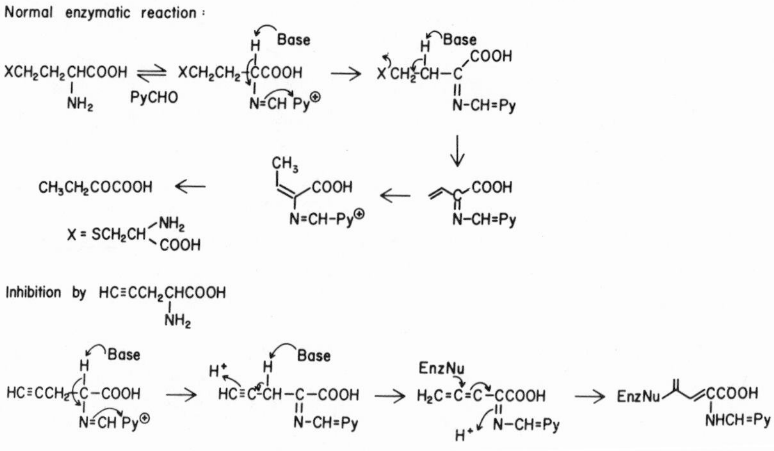

Normal enzymatic reaction :

$XCH_2CH_2CHCOOH \rightleftharpoons XCH_2CH_2CCOOH \longrightarrow X CH_2CH-C$
       | | COOH
      NH_2 PyCHO $N=CH$ $Py^{\oplus}$ $N-CH=Py$

$CH_3CH_2COCOOH \longleftarrow$ CH_3 $COOH \longleftarrow$ $COOH$

                          $N=CH-Py^{\oplus}$ $N-CH=Py$

$X = SCH_2CH$ $^{NH_2}_{COOH}$

Inhibition by $HC\equiv CCH_2CHCOOH$
                          |
                         NH_2

$HC\equiv CCH_2C-COOH \rightarrow HC\equiv C-CH-C-COOH \rightarrow H_2C=C=C-CCOOH \rightarrow EnzNu \quad CCOOH$

$N=CH$ $Py^{\oplus}$ $N-CH=Py$ $N-CH=Py$ $NHCH=Py$

Figure 2.   Inhibition of γ-cystathionase

Abeles and Maycock (2) and Walsh (5) as well as to a symposium
proceedings which offers the most recent and comprehensive reviews
in the area (6). This review will restrict itself to the irreversible inhibition of pyridoxal phosphate (PyCHO)-dependent enzymes,
a class of enzymes which has proven to be generally susceptible to
inhibition by suicide enzyme inactivators.

γ-Cystathionase, which catalyzes the reaction shown in Fig. 2,
has the ability to catalyze the abstraction of both the α- and β-
protons of the substrate. In an early example of inhibition of a
PyCHO-dependent enzyme <u>via</u> the suicide concept, Abeles and Walsh
(7) demonstrated that propargylglycine (II) is an irreversible
inhibitor of γ-cystathionase, which accepts it as a substrate.
The normal proton abstraction which precedes β-elimination induces
allene formation from II. The allene, being a Michael acceptor
then inactivates the enzyme <u>via</u> an alkylative process involving a
nucleophilic (Nu) residue on the enzyme. Propargylglycine has
since been shown to also irreversibly inactivate glutamate-pyru-
vate transaminase (8).

Because of its physiological importance, γ-aminobutyric acid
transaminase (GABA-T), the PyCHO-dependent enzyme responsible for
the catabolism of the inhibitory neurotransmitter, γ-aminobutyric
acid (GABA) (Fig. 3), has been subjected to a variety of
approaches for inhibition by suicide enzyme inactivators (9).
Ethanolamine-0-sulfate (III; Fig. 4) was the first rationally
designed irreversible inhibitor of GABA-T (10). III, being
accepted as a substrate in the same manner as is GABA, forms a
Schiff base with PyCHO. In this way, the adjacent C-H bond is
activated so that proton abstraction by the enzyme is facilitated.
The resulting carbanion then induces elimination of sulfate and
thereby generates an α,β-unsaturated imine which alkylates a
nucleophilic (Nu) residue in the active site (Fig. 4). Unfortu-
nately, III does not readily penetrate the blood-brain barrier and
its use as a tool to study GABA function has been limited.

As GABA-T operates by Schiff's base-mediated proton abstraction, γ-acetylenic GABA (IV; Fig. 5), a substrate analogue bearing an acetylenic function attached to the γ-carbon atom could by
analogy to the inhibition of β-hydroxydecanoylthioester dehydrase
by acetylenic substrate analogues (4), irreversibly inhibit this
enzyme (Fig. 5). Thus, reprotonation of the enzymatically-gen-
erated propargylic carbanion could lead to allene formation. As
such an allene, being conjugated to the imine function, would be
an alkylating agent irreversible inhibition should ensue.

Based on this premise, γ-acetylenic GABA (IV) was synthesized (11) and found to be an irreversible inhibitor of GABA-T,
<u>in vitro</u> and <u>in vivo</u> (12). Thus, when GABA-T, partially purified
from pig brain, is incubated for varying time periods with γ-
acetylenic GABA, a time-dependent inactivation process is observed
which follows pseudo first-order kinetics. Enzyme half lives
range from 28 minutes to 9 minutes with concentrations of inhibitor between 0.029 mM and 0.29 mM. Time dependent inactivation is

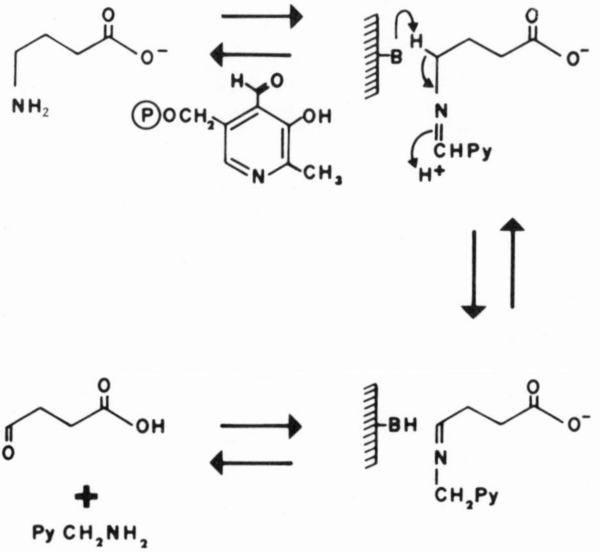

Figure 3.　Mechanism of γ-aminobutyric acid transaminase

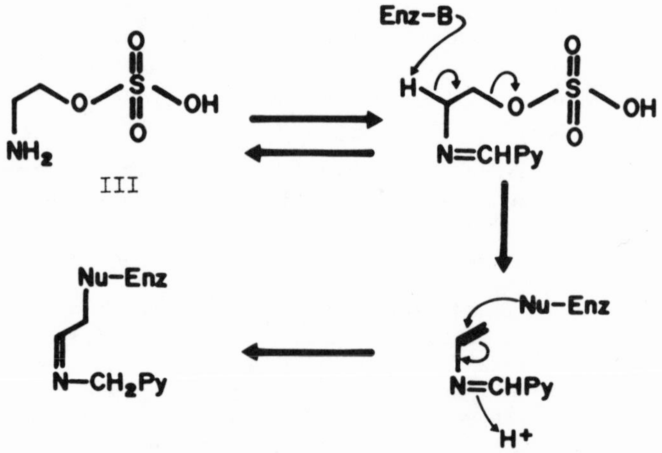

Figure 4.　Inhibition of γ-aminobutyric acid transaminase by ethanolamine-0-sulfate

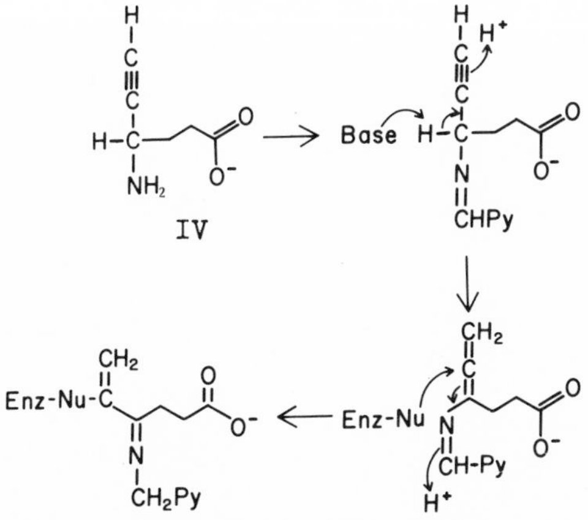

Figure 5.   Inhibition of γ-aminobutyric acid transaminase by 4-aminohex-5-ynoic acid (γ-acetylenic GABA)

indicative that covalent modification has occurred (2) and this is
confirmed by the finding that prolonged dialysis of inhibited
preparations against several buffers containing PyCHO does not
restore enzyme activity. That the loss of enzyme activity is
first order at constant inactivator concentration is evidence that
inactivation occurs before the inactivator is released from the
enzyme. When GABA is added to the incubation medium, the rate of
inactivation induced by γ-acetylenic GABA is dramatically reduced.
However, when α-ketoglutarate is also present, this protection
against inactivation is lost. Evidently, GABA is able to protect
the enzyme against inhibition because the holoenzyme is transformed, in one turnover, to the pyridoxamine form. This cannot
bind the inhibitor as Schiff's base formation is no longer possible. In the presence of α-ketoglutarate, the pyridoxal form is
regenerated and inhibition can ensue. γ-Acetylenic GABA thus
appears to be a suicide inactivator of GABA-T as most of the
kinetic criteria (2) which are indicative of substrate-induced
irreversible enzyme inactivation are satisfied.

    Inhibition of PyCHO-dependent enzymes by β,γ-unsaturated
amines is not limited to exploitation of carbanion-induced
acetylene-allene isomerism. As demonstrated in Fig. 6, allyl
amines can also irreversibly inactivate PyCHO-dependent enzymes
via mechanisms involving double bond isomerism. Thus, if γ-vinyl
GABA (V) were a substrate for GABA-T, the normal transamination
mechanism (path a) would lead to a conjugated imine. Alternatively (path b), isomerism of the double bond would generate a new
double bond, which would be conjugated through to the pyridine
ring. In either case, an alkylating agent would be formed as a
result of the enzyme's own mode of action. The transamination
pathway is the one which has been found by Rando et al. (13) to
be operative in the irreversible inhibition of aspartate aminotransferase by 2-amino-4-methoxy-trans-3-butenoic acid. On the
other hand, α-vinyl glycine inhibits the same enzyme via the
isomerism pathway (14). As anticipated, γ-vinyl GABA (V) is an
irreversible inhibitor of GABA-T in vitro (15) and in vivo (16)
although whether inhibition occurs via transamination (path a) or
isomerism (path b) is as yet unknown.

    In principle, enzyme inhibitors which require transformation
by the target enzyme prior to that enzyme's irreversible inhibition should be extremely specific as they should inhibit only
those enzymes which can accept them as substrates. In keeping
with this, γ-acetylenic GABA (IV) has little effect on alanine and
asparate aminotransferases. However, it has now been found to be
an irreversible inhibitor of the PyCHO-dependent glutamic acid
decarboxylase (GAD) (17) and ornithine aminotransferase (OAT) (18).
Since GABA is a substrate for OAT, inhibition of this enzyme by
γ-acetylenic GABA (IV) is not surprising. Inhibition of GAD by
IV was unexpected and will be discussed later. γ-Vinyl GABA (V)
on the other hand, appears to be the most specific inhibitor of
GABA-T known. To date, no other enzyme has been found to be

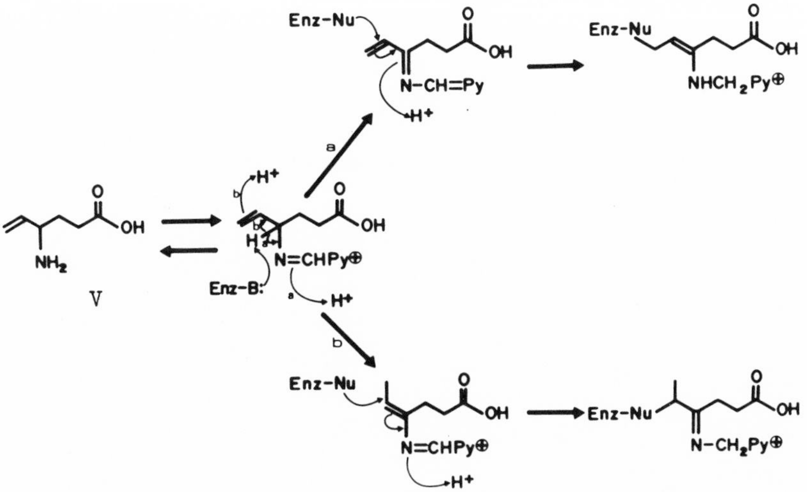

Elsevier/North Holland Biomedical Press

Figure 6. Inhibition of γ-aminobutyric acid transaminase by 4-aminohex-5-enoic acid (γ-vinyl GABA) appreciably inhibited by this compound (18).

In each of the examples of suicide inactivation thus far
discussed, inhibition relies on the addition of an appropriately-
positioned active-site nucleophilic residue to an electrophilic
species which has been generated as a result of the enzyme's usual
mechanism of action. Nature has recently provided a clue as to
how to avoid the dependency on fortuitous positioning of a suit-
ably-placed nucleophile in the enzyme active site, necessary if
irreversible inhibition is to ensue via an alkylation route.
Kobayashi et al. (19) reported the isolation of another GABA-T
inhibitor, gabaculine (VI; Fig. 7) from a Streptomyces species.
Rando (20) has subsequently demonstrated that the mechanism of
irreversible inhibition of GABA-T by gabaculine does not involve
an alkylation process but instead results from the covalent linkage of the transformed inhibitor to the co-enzyme, the driving
force being aromatization of the cyclohexadiene unit (Fig. 7).
A synthetic isomer of gabaculine (VII; Fig. 8) has similar biochemical activity in vitro and in vivo (21) to gabaculine (22).
The remaining isomer of gabaculine (VIII; Fig. 8) has recently
been synthesized and awaits testing (23).

Reminiscent of the irreversible inhibition of GABA-T by
ethanolamine-O-sulfate (10), which involves enzyme-induced β
elimination of sulfate to generate an electrophilic Michael
acceptor, β-haloamino acids have been found to lead to irreversible inhibition via β-elimination mechanisms. Thus bacterial
alanine racemase is irreversibly inhibited by β-chloro-D-alanine
(24), β-fluoroalanine (25) and by β,β,β-trifluoroalanine (26).
β,β,β-Trifluoroalanine has also been found to be an irreversible
inactivator of γ-cystathionase (26, 27), the enzyme previously
shown to be inactivated by propargylglycine (7).

The concept of inhibition via β elimination of fluoride ion
has now been extended to the irreversible inhibition of α-amino
acid decarboxylases. Ornithine decarboxylase (ODC), which catalyzes the decarboxylation of ornithine to putrescine is irreversibly inhibited by α-difluoromethylornithine (IX; Fig. 9) (28). In
this case, the carbanion formation which precedes β elimination is
generated by loss of CO_2, and not by proton abstraction (Fig. 9).
Similarly, aromatic amino acid decarboxylase is irreversibly
inhibited by α-difluoromethyl-3,4-dihydroxyphenylalanine (29)
while histidine decarboxylase, ornithine decarboxylase and aromatic
amino acid decarboxylase have been inhibited by the corresponding
α-monofluoromethylamino acids, respectively (29).

As suicide enzyme inactivators must be substrates for the
target enzyme, they have generally been designed by incorporating
a latent-reactive functionality into the structure of the enzyme's
natural substrate. However, in view of the microscopic reversibility principle they may conceptually be analogues of the product. Thus, in considering the decarboxylation of glutamic acid to
GABA by GAD, if γ-acetylenic GABA (IV) can replace the product of
decarboxylation GABA in the active site, the proton abstraction

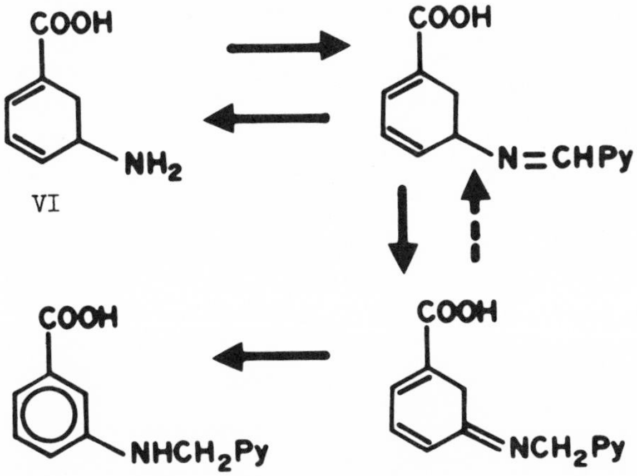

Figure 7.   Inhibition of γ-aminobutyric acid transaminase by gabaculine

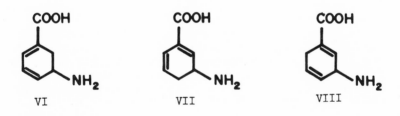

GABACULINE

Figure 8.   Isomers of gabaculine

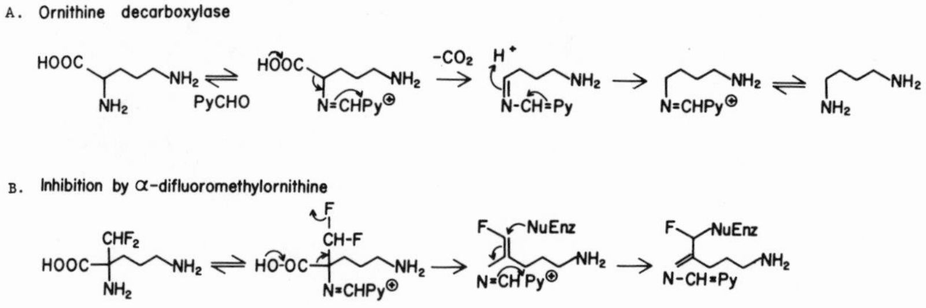

A. Ornithine decarboxylase

B. Inhibition by α-difluoromethylornithine

IX

Figure 9. Inhibition of ornithine decarboxylase by α-difluoromethylornithine

implicit in the reverse direction would lead to propargylic anion
formation and hence irreversible inactivation in a similar manner
to that proposed for the inhibition of GABA-T by this compound
(Fig. 5). For bacterial GAD this argument is supported by the
absolute stereochemistry of the inactivation process. As GAD-
catalyzed replacement of COOH in 2-S-glutamic acid by H occurs
with retention of configuration (31), it is the pro R hydrogen in
GABA which is potentially labile in the reverse direction. That
bacterial GAD is inhibited by 4(R)-4-aminohex-5-ynoic acid ((-)-
γ-acetylenic GABA), is in agreement with the expected stereochemistry of proton abstraction (17).

The concept of irreversible inhibition by product analogues
appears to be general as ODC is inhibited by (±)-5-hexyne-1,4-
diamine, the acetylenic analogue of the product of decarboxylation
putrescine (28) while aromatic amino acid decarboxylase and histidine decarboxylase are inhibited by α-monofluoromethyldopamine and
α-monofluoromethylhistamine, respectively (30). Interestingly, it
is 4(S)-4-aminohex-5-ynoic acid which inhibits mammalian GAD (32).
Thus it appears that in some cases a mechanism other than microscopic reversibility is operative for inhibition by product analogues, although the inactivation mechanism still involves enzyme
catalysis.

PyCHO-dependent enzymes which catalyze condensation reactions
have also been found to be inhibited by suicide inactivators.
Thus tryptophan synthetase, which catalyzes the addition of serine
to indole *via* an α,β-unsaturated imine derivative, is inactivated
by α-cyanoglycine (33). In this case, α-cyanoglycine, an analogue
of the substrate serine, undergoes Schiff base formation. Proton
abstraction then occurs and the resultant α-cyano carbanion is
apparently reprotonated to generate a reactive keteneimine which
can alkylate a nucleophilic active site residue.

δ-Aminolevulinate synthetase catalyzes the condensation of
the Schiff's base of glycine with succinoyl-CoA. Recently, it has
been found that 2-amino-4-methoxy-*trans*-3-butenoic acid, previously found to inhibit aspartate aminotransferase (13), also
irreversibly inactivates this enzyme (34).

In conclusion, suicide enzyme inactivators offer a powerful
method for the selective irreversible inhibition of enzymes.
Although this review has concentrated on pyridoxal phosphate-
dependent enzymes the approach is also valid for the irreversible
inhibition of other types of enzymes (6) and may offer a means
for the rational design of therapeutically-useful substances.

Acknowledgments

Figures 4, 6, and 7 have been reproduced from B. W. Metcalf
et al., "Enzyme activated irreversible inhibition of trans-
aminases" in "Enzyme-Activated Irreversible Inhibitors,"
N. Seiler, M. J. Jung and J. Koch-Weser eds., 1978, Elsevier/North
Holland Biomedical Press.

Literature Cited

1. Rando, R.R., Science, 1974, 185, 320.
2. Abeles, R.H.; Maycock, A.L.; Acc. Chem. Res., 1976, 9, 313.
3. Endo, K.; Helmkamp, G.M.; Bloch, K., J. Biol. Chem., 1970,
   245, 4293.
4. Wood, B.J.B.; Ingraham, L.L., Nature, 1965, 205, 291.
5. Walsh, C., Horizons Biochem. Biophys., 1977, 3, 36.
6. Seiler, N.; Jung, M.J.; Koch-Weser, J., eds., "Enzyme-
   Activated Irreversible Inhibitors," 1978, Elsevier/North
   Holland Biomedical Press.
7. Abeles, R.H.; Walsh, C.T., J. Amer. Chem. Soc., 1973, 95,
   6124.
8. Marcotte, P.; Walsh, C., Biochem. Biophys. Res. Commun.,
   1975, 62, 677.
9. Metcalf, B.W., Biochem. Pharm., in press.
10. Fowler, L.J.; John, R.A., Biochem. J., 1972, 130, 569.
11. Metcalf, B.W.; Casara, P., Tetrahedron Letts., 1975, 3337.
12. Jung, M.J.; Lippert, B.; Metcalf, B.W.; Schechter, P.J.;
    Bohlen, P.; Sjoerdsma, A., J. Neurochem., 1977, 28, 717.
13. Rando, R.R.; Relyea, N.; Cheng, L., J. Biol. Chem., 1976,
    251, 3306.
14. Gehring, H.; Rando, R.R.; Christen, P., Biochemistry, 1977,
    16, 4832.
15. Lippert, B.; Metcalf, B.W.; Jung, M.J.; Casara, P., Eur. J.
    Biochem., 1977, 74, 441.
16. Jung, M.J.; Lippert, B.; Metcalf, B.W.; Bohlen, P.;
    Schechter, P.J., J. Neurochem., 1977, 29, 797.
17. Jung, M.J.; Metcalf, B.W.; Lippert, B.; Casara, P., Bio-
    chemistry, 1978, 17, 2628.
18. Jung, M.J.; Seiler, N., J. Biol. Chem., 1978, 253, 7431.
19. Kobayashi, K.; Miyazawa, W.; Terahara, A.; Mishima, H.;
    Kurihara, H., Tetrahedron Lett., 1976, 537.
20. Rando, R.R., Biochemistry, 1977, 16, 4604.
21. Metcalf, B.W.; Jung, M.J., Molecular Pharm., in press.
22. Rando, R.R.; Bangerter, F.W., Biochem. Biophys. Res. Commun.,
    1977, 76, 1276.
23. Danishefsky, S.; Hershenson, F.M., J. Org. Chem., 1979, 44,
    1180.
24. Manning, J.M.; Merrifield, N.E.; Jones, W.M.; Gotschlich,
    E.C., Proc. Nat. Acad. Sci., U.S.A., 1974, 71, 417.
25. Wang, E.; Walsh, C., Biochemistry, 1978, 17, 1313.
26. Silverman, R.B.; Abeles, R.H., Biochemistry, 1976, 15, 4718.
27. Silverman, R.B.; Abeles, R.H., Biochemistry, 1977, 16, 5515.
28. Metcalf, B.W.; Bey, P.; Danzin, C.; Jung, M.J.; Casara, P.;
    Vevert, J.P., J. Amer. Chem. Soc., 1978, 100, 2551.
29. Palfreyman, M.G.; Danzin, C.; Bey, P.; Jung, M.J.; Ribereau-
    Gayon, G.; Aubrey, M.; Vevert, J.P.; Sjoerdsma, A., J.
    Neurochem., 1978, 31, 927.

30. Kollonitsch, J.; Patchett, A.A.; Marburg, S.; Maycock, A.L.;
    Perkins, L.M.; Doldouras, G.A.; Duggan, D.E.; Aster, S.A.,
    Nature, 1978, 274, 906.
31. Yamada, Y.; O'Leary, M.H., Biochemistry, 1978, 17, 669.
32. Bouclier, M.; Jung, M.J.; Lippert, B., Eur. J. Biochem., in
    press.
33. Miles, E.M., Biochem. Biophys. Res. Commun., 1975, 64, 2118.
34. Dashman, T.; Kamm, T.T., Life Sciences, 1979, 24, 1185.

RECEIVED October 24, 1979.

INDEX

A

γ-Acetylenic GABA, inhibition of
γ-aminobutyric acid transami-
nase by 4-aminohex-5-ynoic acid ... 245*f*
γ-Acetylenic GABA irreversible
inhibitor
of GABA-T 243
of glutamic acid decarboxylase
(GAD) 246
of ornithine aminotransferase
(OAT) 246
changes during frozen storage 106
filaments of carp before and after
frozen storage, F- 108*f*
properties 97
Activation of pro-histidine decar-
boxylase 53*f*
Active center of enzyme definitions .. 26
Active-site selective reagents 26–28, 27*t*
classification 26–28
after frozen storage, changes in
solubility of carp 110*f*
before and after frozen storage,
carp 101*f*
changes during frozen storage 100
aggregation 100
ATPase activity decrease in 102
dissociation into actin and
myosin 100–102
during frozen storage, aggregation
of fish 101*f*
properties 98
Acylation of free α-NH₂ groups of
proteins 54
at different pH values, rates of 158*t*
effect of temperature on 158*t*
of lysine to dehydroalanine, effect
of calcium chloride concentra-
tion on initial rates of 154*t*
Adenosinediphosphate (ADP) 55
(Adenosinediphosphate–ribose) protein, formation of poly- 56*f*
Adenosinediphosphate ribosylated
nuclear proteins, biological
function of 57
Adenosinediphosphate ribosylation .. 5–57
of protein 56*f*
ADP (*see* Adenosinediphosphate) 55
Affinity labeling of a reactive site 29*f*

Affinity reagent(s) 28, 31
of fish acto-myosin during frozen
storage 101*f*
of myosin, crosslinkages in 107–109
of myosin and L-meromyosin 112–114
Aldehydes, soybean odor from volatile 197
Aldehydes with soy protein products,
interaction of volatile 195–196
1-hexanal with partially denatured
soy protein, binding constant
(K) for 197
protein, effect of denaturation of 196–197
Alkali
-induced effects of racemization
and lysinoalanine formation,
discrimination between 178
on inhibitory activity, effect of
disulfide bond modification by 35*f*
on proteins, effects of 16–21
-treated food proteins, amino acid
racemization in 165–186
lysozyme 21*t*
proteins
amino acid formation in 156*f*
deamidation in 173
factors affecting biological
response to 179–180
soy proteins, cytotoxic and therapeutic consequences of 178
and untreated proteins, enantiomeric ratios in 170*t*
treatment
on food proteins, effects of 165
β-elimination scheme for disul-
fides on 20*f*
of proteins, significance and
application of 159–160
food processing 159–160
protein solubility changes 159
texturizing of foods 159
Alkaline
conditions on proteins, adverse
effects of 145–146
conditions in treatment of pro-
teins uses of 145
solution, protein changes in 145–160
solution, protein reactions in 146
denatuartion 146–147
α-elimination mechanism 147, 150
β-elimination reaction, effect of
conditions on 151–155

Alkaline (*continued*)
  solution, protein reactions in (*continued*)
    hydrolysis 147
      arginine to ornithine 147
      peptide and amide bonds 147
      rates of racemization 148–150
Allylamines inactivation of PyCHO-
  dependent enzymes 246
Alterations of the polyamino acid
  chains, classification, covalent 50
Amine with carbonyl group, reaction
  mechanism of a strongly basic 13*f*
Amino acid(s)
  chains, classification, covalent
    alterations of the poly- 50
  composition
    matrix collagen 67*t*
    mature elastin 67*t*
    microfibrillar protein 67*t*
    of tropoelastin and precursor 71*t*
  decarboxylases, irreversible inhi-
    bition of α- 248
  in deteriorations, chemical reac-
    tions of 6–12
  experimental procedures, racemi-
    zation of 166–169
  formation in alkali-treated proteins 156
  human utilization of D-enantio-
    mers of essential 182–183
  in proteins with peroxide, oxida-
    tions of 11*f*
  racemization
    in alkali-treated food proteins ..165–186
    first-order kinetic equations for
      reversible 185–186
    of protein-bound 169–177
  residues
    modification of C-terminal 52–54
    modification of N-terminal 52–54
    in proteins, racemization of 149*t*
  sequences in non-crosslinked elastin,
    common 68*t*
  side chains, modification of 21, 54–58
  side chains in proteins, dye-cata-
    lyzed photooxidation of 22*f*
γ-Aminobutyric acid transaminase
  γ-acetylenic GABA irreversible
    inhibitor of 243
  by 4-aminohex-5-enoic acid γ-
    vinyl GABA), inhibition of 247*f*
  by 4-aminohex-5-ynoic acid (γ-
    acetylenic GABA), inhibition
    of 245*f*
  by ethanolamine-O-sulfate, inhibi-
    tion of 244*f*

γ-Aminobuteric acid transaminase
  (*continued*)
  ethanolamine-O-sulfate irreversi-
    ble inhibitor of 243
  by gabaculine, inhibition of 249*f*
  gabaculine irreversible inhibition of 248
  mechanism of 244*f*
  γ-vinyl GABA irreversible inhibitor
    of 246
Aminohex-5-enoic acid (γ-vinyl
  GABA), inhibition of γ-amino-
  butyric acid transaminase by
  4- 245*f*, 247*f*
Amylase, and its dextran conjugate,
  heat inactivation (60°C) of β- 132*f*
Antioxidative reaction pathways 84–86
Asn-linked glycoproteins, specificity of 57
Aspartic acid
  content in commercial food prod-
    ucts, D- 184*t*
  racemization in casein, effect of
    protein concentration on 181*t*
  racemization in modified casein 181*t*
Autolysis of trypsin and trypsin–
  dextran conjugate 132*f*
Autoxidation followed by decomposi-
  tion, protein-bound lipids, prob-
  lems from 200–201

B

Binding of 1-hexanal by soy protein .. 198*f*
Binding of unwanted compounds, deterioration of food proteins
  by 195–207
Biological
  oxidation–reduction reactions,
    flavins, cofactors in 83
  oxidation–reduction reactions,
    hemes, cofactors in 83
  response to alkali-treated proteins,
    factors affecting 179–180

C

C-terminal amino acid residues,
  modification of 52–54
Carbohydrate stabilization of enzymes 125
Carbonyl group, reaction mechanism
  of a strongly basic amine with 13*f*
Carcinogenicity of fluorescent compounds 204–206
Carp
  actomyosin after frozen storage,
    changes in solubility of 110*f*
  actomyosin before and after
    frozen storage 101*f*

Carp (*continued*)
before and after frozen storage,
F-actin filaments of 108*f*
L-meromyosin before and after
frozen storage, reconstituted
paracrystals of 105*f*
myosin before and after frozen
storage, reconstituted spindle-
shaped filaments of 105*f*
Casein
aspartic acid racemization in
modified 181*t*
browning, effect of phenolic com-
pounds on 205*t*
effect of protein concentration on
aspartic acid racemization in .. 181*t*
Catalase photoinactivation 89–91
of mitochondrial fraction 90*t*
Cellular systems, prevention of oxidative damage in 84
Chemical
changes in elastin as function of
maturation 63–92
deterioration of muscle proteins
during frozen storage 95–117
deteriorations in derivatization of
proteins 28–34
deteriorative changes of proteins,
overview on 1–44
modification of proteins, posttranslational 49–60
modifications, toxic compounds produced in foods and feeds by 25*t*
Circular dichroism spectra for
ovotransferrin 7*f*
Collagen, amino acid composition
matrix 67*t*
Conformation, protein 16
Conformational properties, ovotransferrin 6
Conjugate(s)
heat inactivation (60°C) of β-
amylase and its dextran 132*f*
nature of soluble dextran–
enzyme 139–141
by pepsin, inactivation of ribonuclease and ribonuclease–
dextran 134*f*
preparation of polysaccharide–
enzyme 130*t*
properties of synthetic dextran–
enzyme
carbohydrate on enzyme–substrate interaction, effect of .. 133
conjugate molecules, effect of
size on properties 139
conjugated enzyme, effect of pH 138
conjugation, "non-effects" of 139
dextranase treatment 137

Conjugate(s) (*continued*)
properties of synthetic dextran–enzyme
(*continued*)
enzyme inhibitors, effect of 135
heat stability 129–131
protein denaturants, effect of ..131–133
proteolytic degradation 131
removal of metal–ion cofactors .. 131
specificity and action pattern ..137–138
synthesis of soluble dextran–
enzyme 125–129
synthetic dextran–enzyme 129–139
trypsin–dextran 140*f*
autolysis of trypsin and 132*f*
inactivation of trypsin and 134*f*
by ovomucoid, inhibition of
trypsin and 136*f*
by protease inhibitors, inhibi-
tion of native trypsin and 137*t*
Conjugated enzyme preparations,
insolubilization of 128
with dextran, preservation of
enzymes by 125–142
of enzyme and soluble polysaccharide, conditions 127–129
loss of enzymatic activity during 126
tests for extent of enzyme–
dextran 128–129
Copper, role of 72
Covalent alterations of the poly-
amino acid chains, classification .. 50
Crosslink formation 72
Cross-linkages in aggregation of
myosin 107–109
Cross-linkages in proteins during
heating, formation of amide 13*f*
Cross-linking of RNase A, inacti-
vation and 17*f*
Cryoprotectants
dianionic 116*f*
for fish muscle proteins 111
as water structure modifiers 115
Cryoprotective agents, effect of 109–111
Cryoprotective effect(s)
for fish muscle proteins, requirements for 111
on freeze denaturation of protein,
sodium glutamate 106, 107, 111
mechanism of 114–115
α-Crystallin, mechanism of N-acetylation of 54
γ-Cystathionase, γ- 248
inhibition of 242*f*
propargylglycine (II) irreversible
inhibitor of 243
Cysteine, dichlorovinyl- 24
Cytotoxic and therapeutic consequences of alkali-treated soy
proteins 178–182

D

Damage
  characterization of pattern of
    intracellular 86
  to mammalian cells by visible light,
    photooxidative 83
  in mitochondria, possible pathways
    of oxidative 85f
  pathways, photooxidative 91–92
  to proteins by visible light, photo-
    oxidative 83
Damaging effects of visible light exposure, cultured mammalian cells .. 86
Deamidation in alkali-treated proteins 173
Definitions, active center of enzyme .. 26
Demosine from lysine, scheme
  formation of 76f
Denaturation 2–6
  constants, transition-state 5t
  globular proteins 3
  during frozen storage 116f
  unfolding in freeze 114
  of α-helical proteins during frozen
    storage 113f
  mechanism of freeze 111–114
  of muscle proteins 95
  freeze 98–109
  of myosins 99f
  of ovotransferrin 8t–9t
  of protein, sodium glutamate, cryo-
    protective effect on freeze 107, 111
  relationship to native protein
    structures 3
Dephosphorylation, regulation of
  enzymatic activity by 55
Derivatization of protein, chemical
  deteriorations in 28–34
Desmosines, formation of 73
Detections of protein deteriorations .. 39t
Deteriorated proteins, detection and
  determination of 37–42
  methods 38–42
  problems 37–38
Deterioration(s)
  areas for future investigations
    on protein 42–45
    foods and feeds 43
    living systems 43
  methods for detection, charac-
    terization, quantitation,
    purification 44
  chemical reactions of amino acids
    in 6–12
  detection of protein 39t
  in derivatization of proteins,
    chemical 28–34
  of food proteins by binding of
    unwanted compounds 195–207
  involving disulfide linkages 16

Deterioration (continued)
  involving lysines 12–16
  of muscle proteins during frozen
    storage 95–117
  occurrence of protein 2
  reversibility of proteins 34–37
Deteriorative change(s)
  in protein during frozen storage in
    food production, use of 227
  of proteins
    during soybean food
      processing 211–239
    overview on chemical 1–44
    use in foods 211–239
  of soybean protein during heating
    on enzyme digestibility, effect
      of 234
  in soy milk protein by water
    evaporation 219
Deteriorative reactions, beneficial
  effects 10–12
Dextran
  activation of soluble 126
  conjugate
    heat inactivation (60°C) of
      β-amylase and its 132f
    by pepsin, inactivation of ribo-
      nuclease and ribonuclease- .. 134f
    trypsin- 140f
    by protease inhibitors, inhibi-
      tion of native trypsin and .. 137t
  of trypsin
    autolysis 132f
    inactivation 134f
    by ovomucoid, inhibition 136f
  conjugation, tests for extent of
    enzyme- 128–129
  –enzyme conjugates
    nature of soluble 139–141
    properties of synthetic 129–139
    carbohydrate on enzyme–sub-
      strate interaction, effect of 133
    conjugate molecules, effect of
      size on properties 139
    conjugated enzyme, effect of
      pH 138
    conjugation, "non-effects" of 139
    dextranase treatment ...129–131, 137
    enzyme inhibitors, effect of 135
    protein denaturants, effect
      of 131–133
    proteolytic degradation 131
    removal of metal–ion cofactors 131
    specificity and action
      pattern 137–138
    synthesis of soluble 125–129
  mechanism of enzyme stabilization
    by 141
  preservation of enzymes by con-
    jugation with 125–142

α-Difluoromethylornithine, inhibition
of ornithine decarboxylase by 250f
Displacement of an aromatic sulfonate 30f
Disulfide(s)
on alkali treatment, β-elimination
scheme for 20f
bond modification by alkali on inhibitory activity, effect of 35f
hydrolytic scissions of 16
linkages, deteriorations involving 16
Drying
effect of heating conditions of soy
milk before 212
effect of heating of soy milk
before 213f
insolubilization of soy milk
protein(s) 214f, 217f
irreversible insolubilization of
soybean protein during 211–219
mechanism of insolubilization of
heated soy milk during 212

E

Egg(s)
effect of adding dilute thioglycol
to broken-out 18f
Maillard reaction in dried whole 14
starch-gel electrophoretic patterns
of incubated infertile 15f
white ovotransferrin, chicken 21
Elastic fiber alterations during
maturation 75–80
Elastin 63
amino acid composition mature 67t
biosynthesis of 65–71
common amino acid sequences in
non-crosslinked 68t
elasticity, models 64–65
fibers, alterations in diseased or
aged 80
fibers, synthesis of mature 70f
fibrils, formation of stable 71–75
fibrils, lysyl oxidase, role in
stabilizing 72–73
form non-crosslinked 69
forms of soluble 69
function 64
as function of maturation, chemical
changes in 63–92
index of maturation 77
isolation, procedural difficulties 75
-like proteins secretion 65
metabolic turnover of 77
model 66f
structure models 64
synthesis in tissues, stimulating
factors of 69–71
terminology 63
elastic fiber 64

Elastin (continued)
terminology (continued)
non-crosslinked 63
tropoelastin 63
turnover of arterial 79f
Elastolysis 80
Electropulse column(s) 283, 293f, 386
β-Elimination
effect of calcium ion on intitial
rate of 152t
effect of hydroxide ion concentra-
tion on initial rate of 152t
of phosvitin, effect of calcium chloride concentration on initial
rates of 154t
reaction in proteins, effect of temperature on 153t
scheme for disulfides on alkali
treatment 20f
Enantiomeric ratios in alkali-treated
and untreated proteins 170t
Enantiomers of essential amino
acids, human utilization of D- 182
Enzymatic
activity by dephosphorylation,
regulation of 55
activity by phosphorylation 55
hydrolysis of denatured proteins 237f
hydrolysis of proteins, influencing
factors 234
Enzyme(s)
association equilibrium constants
for 32t
carbohydrate stabilization of 125
changes during frozen storage 107
conjugates
nature of soluble dextran– 139–141
preparation of polysaccharide- 130t
properties of synthetic dextran–
carbohydrate on enzyme–substrate interaction, effect of 133
conjugate molecules, effect of
size on properties 139
conjugated enzyme, effect of
pH 138
conjugation, "non-effects" of 139
enzyme inhibitors, effect of 135
heat stability 129–131
protein denaturants, effect
of 131–133
proteolytic degradation 131
removal of metal–ion cofactors 131
specificity and action
pattern 137–138
conjugates, synthesis of soluble
dextran– 125–129
by conjugation with dextran,
preservation of 125–142
definitions, active center of 26

Enzyme(s) (continued)
–dextran conjugation, tests for
    extent of 128–129
digestibility and nutritive value
    of protein 236–239
inactivators, suicide 241–251
preparations, insolubilization of
    conjugated 128
pyridoxal phosphate (PyCHO)-
    dependent 243
stabilization by dextran,
    mechanism of 141
and soluble polysaccharide, cond-
    ditions, conjugation of 127–129
Enzymic activity during conjuga-
    tion, loss of 126
2',3'-Epoxypropyl β-glycoside of di-
    (N-acetyl-D-glycosamine) 33f
Ethanolamine-O-sulfate, inhibition of
    γ-aminobutyric acid transami-
    nase by 244f
Ethanolamine-O-sulfate irreversible
    inhibitor of GABA-T 243
Extraction of proteins from green
    leaves and algae, problems of 203

F

Feeds by chemical modifications,
    toxic compounds produced in
    foods and 25t
Feeds, undesirable chemical products
    formed in processing 24–26
Fish
    acto-myosin during frozen storage,
        aggregation of 101f
    muscle proteins, cryoprotectants 111
    muscle proteins, requirements for
        cryoprotective effects for 111
Flavins, cofactors in biological oxi-
    dation–reduction reactions 83
Flavin dehydrogenases, photooxi-
    dative attack on 89
Flavors, interaction of proteins
    with 195–199
Fluorescent compounds, carcino-
    genicity of 204–206
Fluorescent compounds, protein
    interaction with 204
Food(s)
    and feeds by chemical modifica-
        tions, toxic compounds pro-
        duced in 25t
    processing, deteriorative changes
        of proteins during soybean ..211–239
    products, D-aspartic acid content
        in commercial 184t
    production, use of deteriorative
        changes in protein during
        frozen storage in 227

Food(s) (continued)
proteins
    amino acid racemization in
        alkali-treated 165–186
    by binding of unwanted com-
        pounds, deterioration of ..195–207
    effects of alkali treatment on 165
    effects of heat treatment on 165
    radical-induced damage in pres-
        ence of unsaturated lipids 202
    reversible and irreversible insolu-
        bilization of soybean protein
        use for 227–233
    undesirable chemical products
        formed in processing 24–26
Formation–deformation of ternary
    protein–oil phosphtidylcholine
    complex 201f
Formation of poly-(ADP-ribose)
    protein 56f
Freeze denaturation
    globular proteins unfolding in 114
    mechanism of 114–115
    of muscle proteins 98–109
    of protein, sodium glutamate,
        cryoprotective effect on 102, 104,
                                            106, 107, 111
Freezing, irreversible insolubilization
    of soybean protein after 219–226
Frozen storage
    actin, changes during 106
    F-actin filaments of carp before
        and after 108f
    actomyosin changes during 100
    aggregation 100
    ATPase activity decreases in 102
    dissociation into actin and
        myosin 100–102
    insolubilization 102
    aggregation of fish acto-myosin
        during 101f
    carp actomyosin before and after 101f
    changes in solubility of carp acto-
        myosin after 110f
    chemical deterioration of muscle
        proteins during 95–117
    decrease of lactate dehydrogenase
        activity during 108f
    denaturation of globular proteins
        during 116f
    denaturation of α-helical proteins
        during 113f
    enzymes, changes during 107
    in food production, use of deterio-
        rative changes in protein
        during 227
    insolubilization of soybean
        protein 222f, 228f
    myofibrils, changes during 106

Frozen storage (*continued*)
myosin, changes during 103
aggregation 103
decrease in ATPase activity 103
myosin subunits, changes during 104–106
H-meromyosin 104–106
L-meromyosin 104–106
in native proteins, contribution of
disulfide bonds to insolubili-
zation during 224
rates of insolubilization during 231*f*
reconstituted paracrystals of carp
L-meromyosin before and after 105*f*
reconstituted spindle-shaped filaments of carp myosin before
and after 105*f*
sarcoplasmic proteins, changes
during 107
tropomyosin changes during 106
troponin changes during 106
Frozen stored meats, undesirable
changes 95

G

GABA (*see* γ-Aminobutyric acid
transaminase) 243
Gabaculine
inhibition of γ-aminobutyric acid
transaminase by 249*f*
irreversible inhibition of GABA-T .. 248
isomers of 249*f*
GABA-T (*see* γ-Aminobutyric acid
transaminase) 243
GAD (glutamic acid decar-
boxylase 246, 248
Globular proteins
denaturation 3
during frozen storage, denatura-
tion of 116*f*
unfolding in freeze denaturation 114
Globulin molecules susceptibility to
proteolysis, 11S 235*f*
Gluconic acid, enzymatic oxidation
of glucose to 26
Glucose
to gluconic acid, enzymatic
oxidation of 26
with hemoglobin, reaction of 60*f*
removal on storage-induced
changes, effect of 17*f*
Glutamic acid decarboxylase (GAD),
γ-acetylenic GABA irreversible
inhibitor of 246
Glutamine synthetase, inhibition of 29*f*
Glycoprotein, general scheme for
biosynthesis of Asn-linked 56*f*
Glycoproteins, specificity of Asn-
linked 57
Glycosylation, protein.. 57–58

H

β-Haloamino acids irreversible
inhibitors 248
Heat treatment
on food proteins, effects of 165
of protein on proteolysis rate,
effect of 234
of soybean protein, effect of 235*f*
on soybean protein hydrolysis
patterns, effect of 238*f*
Heated soy milk, effect of addition
of N-ethylmaleimide to 213*f*
Heating
formation of amide cross-linkages
in proteins during 13*f*
methods on protein nutritive value 12*t*
of soy milk before drying, effect of .. 213*f*
α-Helical proteins during frozen storage denaturation of 113*f*
Hemes, cofactors in biological oxidation–reduction reactions 83
Hemoglobin, reaction of glucose with 60*f*
Hepatocytes, enzyme photoinactiva-
tion in isolated 87*f*
Hepatocytes studies, isolated 86
1-Hexanal retention, soy protein relationship between hydrolysis and .. 198*t*
1-Hexanal by soy protein, binding of .. 198*f*
Histidine imidazoles, photooxidation
of 21
Histidine, photooxidation pathways
for 22*f*
Homoserine 31
Hydrolysis
of denatured proteins, enzymatic 237*f*
patterns, effect of heat treatment
on soybean protein 238*f*
of proteins 171
influencing factors enzymatic 234
β-Hydroxydecanoylthioester dehy-
drase, inhibition of 242*f*

I

Inactivation
(60°C) of β-amylase and its
dextran conjugate, heat 132*f*
and cross-linking of RNase A 17*f*
of lysozyme, influence of copper
concentration on 20*f*
of ribonuclease and ribonuclease–
dextran conjugate by pepsin 134*f*
of trypsin and trypsin–dextran
conjugate 134*f*
Inactivators, suicide enzyme 241–251
Inhibition
of α-amino acid decarboxylases,
irreversible 248
of γ-aminobutyric acid trans-
aminase

Inhibition (*continued*)
by 4-aminohex-5-enoic acid
(γ-vinyl GABA) 247*f*
by 4-aminohex-5-ynoic acid
(γ-acetylenic GABA) 245*f*
by ethanolamine-O-sulfate 244*f*
by gabaculine 249*f*
of γ-cystathionase 242*f*
of glutamine synthetase 29*f*
of β-hydroxydecanoylthioester
dehydrase 242*f*
of native trypsin and trypsin–dex-
trans conjugate by protease
inhibitors 137*t*
of ornithine decarboxylase by
α-difluoromethylornithine 250*f*
by product analogs, irreversible 251
of trypsin and trypsin–dextran
conjugate by ovomucoid 136*f*
Inhibitor(s)
of γ-cystathionase, propargylglycine (II) irreversible 243
of GABA-T,
γ-acetylenic GABA irreversible .. 243
ethanolamine-sulfate irreversible 243
γ-vinyl GABA irreversible 246
of glutamic acid decarboxylase
(GAD), γ-acetylenic GABA
irreversible 246
β-haloamino acids irreversible 248
kcat 241
Inhibitory activity, effect of disulfide
bond modification by alkali on .. 35*f*
Insolubilization
of conjugated enzyme preparations 128
during frozen storage in native
proteins, contribution of di-
sulfide bonds to 224
during frozen storage, rates of 231*f*
of heated soy milk during drying,
mechanism of 212
of soybean protein
after freezing, irreversible 219–226
mechanism 221–224
during drying
irreversible 211–219
intermolecular polymeriza-
tion through hydro-
phobic interactions 218
use in food production 219
polymerization by disulfide
bonds 212–215
disulfide bond interchange
reaction 215
frozen storage 222*f*, 228*f*
mechanisms of 229*f*
use for foods 227–233
artificial meat products 232

Insolubilization (*continued*)
of soybean protein (*continued*)
use for foods (*continued*)
irreversible insolubiliza-
tion 230, 232
charge neutralization 230
product dependence on bond
contribution 232
reversible insolubilization 230
soybean protein products
functional properties 230
of soy milk protein(s), drying ..214*f*, 217*f*
Intracellular damage, characterization of pattern of 86
Isomers of gabaculine 249*f*

K

K_cat inhibitors 241
K_cat reagents 28
Kinetics of base-catalyzed racemiza-
tion 171
Kinetic equations for reversible
amino acid racemization,
first-order 185–186
Kori-tofu 228*f*

L

Lactate dehydrogenase activity during frozen storage, decrease of 108*f*
Light
exposure, cultured mammalian
cells, damaging effects of
visible 86
photooxidative damage to mam-
malian cells by visible 83
photooxidative damage to pro-
teins by visible 83
LINCS (local independently nucleated continuous segments) 3
Lipids
from autoxidation 200–201
food proteins, radical-induced
damage in presence of un-
saturated 202
interaction of protein with 200–202
problems from autoxidation followed by decomposition,
protein-bound 200
Lysines, deteriorations involving 12–16
Lysine, scheme for formation of
desmosine from 76*f*
Lysinoalanine formation, discrimination between alkali-induced
effects of racemization and 178
Lysinoalanine, nephrotoxic action
of 178–180

Lysozyme
  alkali-treated 21t
  enzymatic activities and associa-
    tion constants of 34t
  influence of copper concentra-
    tion on inactivation of 20f
Lysyl oxidase 74f
  role in stabilizing elastin fibrils 72–73

M

Maillard reaction 10, 12–16
  in dried whole eggs 14
Mammalian cells, damaging effects of
  visible light exposure 86
Mammalian cells by visible light,
  photooxidative damage to 83
Matrix collagen, amino acid compo-
  sition 67t
Maturation, chemical changes in
  elastin as function of 63–92
Meats, undesirable changes in frozen
  stored 95
Meromyosin, aggregation of myosin
  and L- 112–114
Meromyosin before and after frozen
  storage, reconstituted paracrys-
  tals of carp L- 105f
Methionine, photooxidation path-
  ways for 22f
Methionine sulfoximine, toxicity of 24
Methylases, characterization, pro-
  teins 54–55
Methylation of free α-NH₂ groups
  of proteins 54
Methylesterase catalysis 55
Microfibrillar protein, amino acid
  composition 67t
Mitochondria, isolated 88–89
  absorption spectrum 88–89
  bioenergetic parameters, pattern of
    change after light exposure 88
Mitochondria, possible pathways of
  oxidative damage in 85f
Mitochondrial fraction catalase,
  photoinactivation 90t
Modification(s)
  by alkali on inhibitory activity,
    effect of disulfide bond 35f
  of amino acid side chains 21, 54–58
  of C-terminal amino acid residues ..52–54
  chemical side reactions during
    protein 10t
  environmental factors, protein 58–59
  of N-terminal amino acid residues ..52–54
  by nonenzymatic reactions 58–59
  of proteins, posttranslational
    chemical 49–60

Modification(s) (continued)
  reaction(s)
    biological function of posttrans-
      lational 49–50
    summary, posttranslational 51f
    terminology 49
  toxic compounds produced in foods
    and feeds by chemical 25t
Modifiers, cryoprotectants as water
  structure 115
Muscle 96
  anatomical structure 96
  chemical constitution 96
  protein(s)
    comparative stability of 98
    composition of 97t
    cryoprotectants for fish 111
    denaturation of 95
    during frozen storage, chemical
      deterioration of 95–117
    freeze denaturation of 98–109
    properties of 97
    requirements for cryoprotective
      effects for fish 111
    solubility classification of
      vertebrate 96
Myofibrils, changes during frozen
  storage 106
Myosin(s)
  before and after frozen storage,
    reconstituted spindle-shaped
    filaments of carp 105f
  changes during frozen storage 103
  aggregation 103
  decrease in ATPase activity 103
  crosslinkages in aggregation of 107–109
  denaturation of 99f
  and L-meromyosin, aggregation
    of 112–114
  properties 97
  subunits, changes during frozen
    storage 104–106
  H-meromyosin 104–106
  L-meromyosin 104–106

N

N-acetylation of α-crystallin,
  mechanism of 54
N-terminal amino acid residues,
  modification of 52–54
Nephrocytomagaly 178–182
Nephrotoxic action of lysinoalanine .. 180
α-NH₂ groups of proteins, acylation
  of free 54
α-NH₂ groups of proteins, methyla-
  tion of free 54
Nonenzymatic reactions, modifica-
  tion by 58–59

Nucleated continuous segments, local
    independently (LINCS) 3
Nutritional implications of racemized
    proteins 182
Nutritive value, protein 12t
    enzyme digestibility and 236–239

O

OAT (ornithine aminotransferase) 246
Ornithine aminotransferase (OAT),
    γ-acetylenic GABA irreversible
    inhibitor of 246
Ornithine decarboxylase by α-di-
    fluoromethylornithine, inhibition
    of 250f
Ovomucoid(s)
    avian 31
    cyanolysis-treated penguin 36t
    inhibition of trypsin and trypsin–
        dextran conjugate by 136f
    turkey 16, 19f
    conformational properties 6
    chicken 5–6
        egg white 21
        circular dichroism spectra for 7f
        denaturation of 8t–9t
        renaturation of 8t–9t
        of acid-denatured 7f
Oxidation(s)
    of amino acids in proteins with
        peroxide 11f
    of glucose to gluconic acid,
        enzymatic 26
    –reduction reactions, flavins, co-
        factors in biological 83
    –reduction reactions, hemes, co-
        factors in biological 83
Oxidative damage in cellular systems,
    preventions of 84
Oxidative damage in mitochondria,
    possible pathways of 85f

P

Pepsin, inactivation of ribonuclease
    and ribonuclease dextran conju-
    gate by 134f
Peptidyl lysine, oxidation of 73
Phenolic compounds on casein
    browning, effect of 205t
Phenolic compounds oxidation to
    pigments 204
Phenylalanine, analgesic effect of D- .. 183
Phosphatidyl choline complex, forma-
    tion–deformation of ternary
    protein–oil– 201f
Phosphorylation, regulation of enzy-
    matic activity by 55
Photodestruction of spin signal 89

Photoinactivation
    catalase 89–91
    in isolated hepatocytes, enzyme 87f
    of mitochondrial fraction catalase .. 90t
    of amino acid side chains in pro-
        teins, dye-catalyzed 22f
    of histidine imidazoles 21
    pathways
        for histidine 22f
        for methionine 22f
        for tryptophan 22f
        for tyrosine 22f
    reactions 21–24
Photooxidative attack on flavin
    dehydrogenases 89
Photooxidative
    damage
        to mammalian cells by visible
            light 83
        pathways 91–92
        to proteins by visible light 83
Phytochrome 203
Pigments, interaction of proteins
    with 202–206
Pigments, phenolic compounds oxi-
    dation to 204
Polyphenols 203
Polysaccharide, conditions, conjuga-
    tion of enzyme and soluble 127–129
Polysaccharide–enzyme conjugates,
    preparation of 130t
Posttranslational
    chemical modification of proteins ..49–60
    modification reactions, biological
        function of 49
    modification reaction summary 51f
Preservation of enzymes by conju-
    gation with dextran 125–142
Processing feeds, undesirable chemi-
    cal products formed in 24–26
Processing foods, undesirable chemi-
    cal products formed in 24–26
Product analogs, irreversible inhi-
    bition by 251
Pro-histidine decarboxylase, activa-
    tion of 53f
Prooxidative reaction pathways 83–84
Propargylglycine (II) irreversible
    inhibitor of γ-cystathionase 243
Protease inhibitors, inhibition of
    native trypsin and trypsin–dex-
    trans conjugate by 137t
Protein(s)
    acylation of free α-NH₂ groups of 54
    ADP ribosylation of 56f
    adverse effects of alkaline condi-
        tions on 145–146
    after freezing, irreversible insolu-
        bilization of soybean 219–226

Protein(s) (*continued*)
 amino acid
  composition microfibrillar 67*t*
  formation in alkali-treated 156*f*
  racemization in alkali-treated
   food 165–186
 analysis, tannins used in 203
 binding of 1-hexanal by soy 198*f*
 by binding of unwanted com-
   pounds, deterioration of
   food 195–207
 biological function of ADP ribo-
   sylated nuclear 57
 -bound amino acids, racemization
   of 169–177
 -bound lipids, problems from auto-
   oxidation followed by decomposition 200–201
 changes in alkaline solution 145–160
 changes during frozen storage,
   sarcoplasmic 107
 chemical deteriorations in derivatization of 28–34
 comparative stability of muscle 98
 components of soybean 225*t*
 composition of muscle 97*t*
 concentration on aspartic acid
   racemization in casein, effect
   of 181*t*
 conformation 16
 contribution of disulfide bonds to
   insolubilization during frozen
   storage in native 224
 cryoprotectants for fish muscle 111
 cytotoxic and therapeutic consequences of alkali-treated soy .. 178
 deamidation in alkali-treated 173
 denaturation globular 3
 denaturation of muscle 95
 detection and determination of
   deteriorated 37–42
   methods 38–42
   problems 37–38
 deteriorations
   areas for future investigations
    on 42–45
   foods and feeds 43
   living systems 43
   methods for detection, char-
    acterization, quantitation,
    purification 44
   detections of 39*t*
   occurrence of 2
   reversibility of 34–37
 drying, insolubilization of soy
   milk 214*f*, 217*f*
 during
   drying, irreversible insolbilization
    of soybean 211–219

Protein(s) (*continued*)
 during (*continued*)
  frozen storage
   chemical deterioration of
    muscle 95–117
   denaturation of globular 116*f*
   denaturation of α-helical 113*f*
  in food production, use of
    deteriorative changes in .. 227
  heating on enzyme digestibility,
    effect of deteriorative
    changes of soybean soy
    sauce, yield of 234
  heating, formation of amide
    cross-linkages in 13*f*
  soybean food processing de-
    teriorative changes of 211–239
 dye-catalyzed photooxidation of
   amino acid side chains in 22*f*
 effect(s)
  of alkali on 16–21
  of alkali treatment on food 165
  of heat treatment on food 165
  of temperature on β elimina-
    tion reaction in 153*t*
 enzymatic hydrolysis of denatured .. 237*f*
 enzyme digestibility and nutritive
   value of 236–239
 factors affecting, biological response
   to alkali-treated 179–180
 with flavors, interaction of 195–199
  denaturation of protein, effect
    of 196–197
  food proteins with volatile alde-
    hydes and ketones 196
  gelatin with 2-alkanone 196
  gelatin and nonvolatile flavor
    nucleotides 195
  soy protein with aldehydes 196
  volatile aldehydes and ketones
    food proteins with 196
 folding 4*f*
 formation of poly-(ADP-ribose) 56*f*
 freeze denaturation of muscle 98–109
 frozen storage, insolubilization
   of soybean 222, 228*f*
 glycosylation 57–58
 from green leaves and algae,
   problems of extraction of 203
 hydrolysis of 171
 hydrolysis patterns, effect of heat
   treatment on soybean 238*f*
 influencing factors enzymatic
   hydrolysis of 234
 interaction with fluorescent compounds 204
 with lipids, interaction of 200–202
 lipid complex, disadvantages of
   protein– 200

Protein(s) (*continued*)
with lipids, interaction of (*continued*)
phosphatidylcholine 200
polar lipids 200
protein–lipid complex, disad-
vantages of 200
mechanisms of insolubilization of
soybean 229f
methylases, characterization 54–55
methylation of free α-NH$_2$ groups of   54
modification, chemical side reac-
tions during 10t
modification, environmental factors 58–59
molecule function, energy of 4f
molecules, unfolding of native 216f
nutritional implications of
recemized 182
nutritive value 12t
effects of heating methods on 12t
–oil–phosphatidylcholine complex,
formation–deformation of a
ternary 201f
overview on chemical deteriorative
changes of 1–44
with peroxide, oxidations of amino
acids in 11f
with pigments, interaction of 202–206
posttranslational chemical modification of 49–60
products, interaction of volatile
aldehydes with soy 195–196
properties of muscle 97
on proteolysis rate, effect of heat
treatment of 234
racemization of amino acid residues in 149t
radical-induced damage in presence
of unsaturated lipids, food 202
reactions in alkaline solution 146, 155
addition reaction 155
denaturation 146–147
α-elimination mechanism 150
β-elimination and racemization .. 147
β-elimination reaction effect of
conditions on 151–155
hydrolysis 147
arginine to ornithine 147
peptide and amide bonds 147
rate of addition effect of conditions on 155
rates of racemization 148–150
requirements cryoprotective effects
for fish muscle 111
S–S bonds 18f
secretion, elastin-like 65
significance and application of
alkali treatment of 159–160
food processing 159

Protein(s) (*continued*)
sodium glutamate, cryoprotective
effect on freeze denatura-
tion of 106, 107, 111
solubility classification of vertebrate muscle 96
structure, denaturation relationship to native 3
structure, renaturation relationship to native 3
unfolding in freeze denaturation,
globular 114
use(s)
of alkaline conditions in treatment of 145
in foods, deteriorative changes
of 211–239
for foods, reversible and irreversible insolubilization of
soybean 227–233
by visible light, photooxidative
damage to 83
Proteolysis 50–52
rate, effect of heat treatment of
protein on 234
11S globulin molecules, susceptibility to 235f
PyCHO (pyridoxal phosphate) dependent enzymes 243
Pyridoxal phosphate (PyCHO)-de-
pendent enzymes 243

R

Racemization
activation energies for 174
in alkali-treated food proteins 165–186
of amino acids, experimental
procedures 166–169
alkali treatment 166–168
amino acid analyses 168
of amino acid residues in proteins .. 149t
in casein, effect of protein concentration on aspartic acid 181t
first-order kinetic equations for reversible amino acid 185–186
kinetics of base-catalyzed 171
and lysinoalanine formation, discrimination between alkali-
induced effects of 178
mechanism of 173
in modified casein, aspartic acid 181t
of protein-bound amino acids 169–177
rate(s)
order of initial 172f
pH dependence of 174
Racemized proteins, nutritional implications of 182
Radical-induced damage in presence
of unsaturated lipids, food
proteins 202

Reagent(s)
active-site selective 26–28, 27t
affinity 28, 31
classification, active-site selective ...26–28
kcat 28
suicide 28
Renaturation
of acid-denatured ovotransferrin 7f
of ovotransferrin 8t–9t
relationship to native protein
structure 3
Ribonuclease and ribonuclease–dex-
trans conjugate by pepsin, inac-
tivation of 134f
Ribosylation, ADP 55–57
RNase A, inactivation and cross-
linking of 17f

S

Sarcoplasmic proteins, changes during frozen storage 107
Scissions of disulfides, hydrolytic 16
Serum transferrin, human 5–6, 21
Sodium glutamate, cryoprotective
effect on freeze denaturation of
protein 107, 111
Solubility of carp actomyosin after
frozen storage, changes in 110f
Soybean
food processing, deteriorative
changes of proteins during ..211–239
globulin 226f
odor from volatile aldehydes 197
protein
after freezing, irreversible insolubilization of 219–226
components of 225t
during drying, irreversible insolubilization of 211–219
during heating on enzyme digestibility, effect of deteriorative changes of 234
effect of heat treatment of 235f
frozen storage, insolubilization
of 222f, 228f
hydrolysis patterns, effect of heat
treatment on 238f
mechanisms of insolubilization
of 229f
use for foods, reversible and irreversible insolubilization
of 227–233
Soy milk 211
before drying, effect of heating 213f
conditions of 212

Soy milk (continued)
effect of addition of N-ethylmaleimide to heated 213f
protein(s), drying, insolubilization of 214f, 217f
Soy protein(s)
binding of 1-hexanal by 198f
cytotoxic and therapeutic consequences of alkali-treated 178
products, interaction of volatile
aldehydes with 195–196
relationship between hydrolysis
and 1-hexanal retention 198t
Stabilization of enzymes, carbohydrate 125
Starch-gel electrophoretic patterns
of incubated infertile eggs 15f
Storage-induced changes, effect of
glucose removal on 17f
Suicide enzyme inactivators 241–251
Suicide reagent 28
Sulfhydryl groups 16
Sulfonate, displacement of an aro-
matic 30f
Sulfur, quantitation of 41f

T

Tannins use in protein analysis 203
Therapeutic consequences of alkali-
treated soy proteins, cytotoxic
and 178–182
Thioglycol to broken-out eggs,
effect of adding dilute 18f
Tofu 233f
Toxic compounds produced in foods
and feeds by chemical modifi-
cations 25t
Toxicity of methionine sulfoxamine 24
Transferrin, human serum 5, 21
Transferrin, model for anion- and
iron-binding sites of 23f
Tropoelastin and precursor, amino
acid composition of 71t
Tropomyosin changes during frozen
storage 106
Troponin, changes during frozen
storage 106
Trypsin–dextran conjugate 140f
by protease inhibitors, inhibition
of native trypsin and 137t
Trypsin and trypsin–dextran conjugate
autolysis of 132f
inactivation of 134f
by ovomucoid, inhibition of 136f
Tryptophan, photooxidation path-
ways for 22f
Turkey ovomucoid 19f
Tyrosine, photooxidation path-
ways for 22f

V

Vertebrate muscle proteins, solubility classification of 96
(γ-Vinyl GABA) inhibition of γ-
aminobutyric acid transaminase
by 4-aminohex-5-enoic acid 247f
γ-Vinyl GABA irreversible inhibitor
of GABA-T 246

Visible light
exposure, cultured mammalian
cells damaging effects of 86
photooxidative damage to mammalian cells by 83
photooxidative damage to proteins
by 83